The Rest of the Story

Forgotten Stories of Plant Pathology History

Robert M. Harveson
Panhandle Research and Extension Center
Scottsbluff, Nebraska
Department of Plant Pathology
University of Nebraska–Lincoln

APS
The American Phytopathological Society
St. Paul, Minnesota U.S.A.

This book was edited and formatted by the author.

Library of Congress Control Number: 2025937421
International Standard Book Numbers:
Print: 978-0-89054-724-3
Online: 978-0-89054-725-0
Epub: 978-0-89054-726-7

Printed in the United States of America on acid-free paper.

The American Phytopathological Society
3285 Northwood Circle, Suite 100
St. Paul, Minnesota 55121, U.S.A

For Tammy, Anna, and Abby

"I have long been of the opinion that no man can have full conception of any subject upon which he may specialize until he has studied the history of its development."

—A. H. R. Buller

"Those who do not remember the past are condemed to repeat it."

—G. Santayana

Preface

About 15 years ago, due to my lifelong passion for history, I began thinking about publishing a book on the history of plant pathology. However, I wanted it to be different than a traditional textbook. Instead, I included various stories of plant diseases, their causative pathogens, and how they may have influenced world history. I additionally tried to select diseases that may not have been highly known or remembered, or small anecdotal portions of better-known diseases that may have assisted with a forgotten effect on history. I additionally attempted to minimize technical language so it could be more easily read and understood by potential newcomers to plant pathology. It was published in 2015 by APS PRESS - *The Bacterium of Many Colors*.

I have additionally produced a similar book published in 2020, but focusing on the history of plant pathology in Nebraska. Since I enjoy researching, composing, and sharing this type of information with others, I hunted through older literature to identify additional topics for more stories of plant pathology history. However, I wanted these new articles to differ from the previous two books. I proposed the new concept of short tales written in the style of Paul Harvey's "The Rest of the Story" series and then appear in the APS's monthly newsletter, *Phytopathology News*. I pitched the idea to the newsletter's editor in chief, Kenny Seebold, and he agreed to let me do so.

That was in June 2016, and since that time, I have produced an article every month, greatly exceeding what I had ever expected or planned at the beginning. Last year I was perusing through my records and surprisingly realized that there were well over 80 articles that had been sent to the newsletter. Thus, I decided to assemble all these stories into one source, and APS PRESS has now agreed to publish them into a new book.

There are 48 articles/chapters that are presented here, roughly one half of the total. They are arranged into 3 specific categories–1) New or Significant Discoveries, 2) Diseases Influencing World History, and 3) Biographies. The chapters in the categories are arranged by pathogens involved, and can be read individually without any specific order. There is additionally a plant breeding and two miscellaneous sections that did not fall into any specific sections. I hope readers now and, in the future, will enjoy reading them as much as I have enjoyed devising and sharing them.

Bob Harveson
Scottsbluff, Nebraska
May 2025

Acknowledgments

First and foremost, great thanks to Allison Rickey for the arduous duty of organizing and formatting all figures, text, and modeling of the overall design of this book.

Thanks also to colleagues Kenny Seebold for agreeing to publish these stories in *Phytopathology News*, and Bob Kemerait for allowing me to continue when he took the reins of editor in chief for the newsletter.

Finally, thanks to APS PRESS Senior Editor Gary Vallad, and the three anonymous reviewers for the time in critically reviewing the manuscript. I know how difficult it is to take the time necessary for reading and assessment. Their comments and suggestions have improved the final product.

Table of Contents

New or Significant Discoveries

Biographies

Bacteria

Fungi

Miscellaneous

Front cover, sunflower research field on the Panhandle Research Extension Center in Scottsbluff, NE.
Back cover, (top) dry bean production field in Scotts Bluff County, Nebraska; (bottom) sugar beet production field in Scotts Bluff county in Nebraska.
Courtesy of R. M. Harveson

CHAPTER 1

Improving Pure Culture Methods for Bacteria

The Forgotten Contributions

By the last decade of the 19th century, bacteria were becoming more widely accepted as legitimate plant pathogens. This was due in large part to several landmark scientific discoveries: the development of the compound microscope for viewing objects as small as bacteria; and development of new methods and synthetic media for growth of bacteria in pure culture.

The credit for developing the proper techniques for isolating and growing bacteria in pure culture is usually given to Robert Koch, the German physician who first demonstrated that bacteria were the causative agents for animal diseases like anthrax, tuberculosis, and cholera. His brilliant works built upon and improved the efforts of both previous and contemporary bacteriologists. However, there are two specific individuals associated with Koch's lab who made significant contributions to the concept of purifying bacterial isolations that are largely unknown today or forgotten. Their stories concerning these milestone discoveries are the rest of the story.

Attempts to culture microorganisms began long before the efforts of mid-19th century bacteriologists such as Louis Pasteur and Joseph Lister. Anton van Leeuwenhoek first observed bacteria from plaque on his teeth with his homemade microscopes, and utilized boiled oat extracts for growing these minute organisms in the 1680s. Others began employing beef broth as liquid media or media amended with beef or yeast cell extracts, milk casein, and even urine. Efforts to purify liquid cultures included multiple serial transfers and serial dilutions.

Due to frequent contamination problems with all these liquid media, new methods using solid substrates for media began, including bread, mixtures of starch, flour and egg whites, and potato slices. These new methods greatly improved upon the previous liquid nutrient or meat-based broths, as solid media enabled better separation and identification of different bacteria from mixed cultures based on colony color or morphology, thereby creating "pure" cultures.

However, it was still difficult in recognizing some bacteria growing on opaque substrates. Koch then set out to find a medium possessing all the desired characteristics: sterile, solid, and transparent. Based on reports by several mycologists, he next turned to melted gelatin poured on sterile glass slides kept in bell jars which enhanced the ability to observe bacterial growth.

Although gelatin was a great improvement over previous methods, there were still several major limitations to its optimal usage. Gelatin's physical properties prohibited its practical usage at higher temperatures. Liquefaction of cultures occurred after a few hours at room temperature. Furthermore, many bacteria produced enzymes that broke down the structure of gelatin-based media, also resulting in undesired liquid cultures again.

It was at this time that two virtually unknown people made their landmark contributions. Walther Hesse was a country physician who spent several years in Koch's lab on sabbatical leave learning new techniques for identifying and isolating airborne bacteria. His wife, Fanny Angelina (Lina to her friends) served as a lab technician for her husband. In 1882, she suggested the use of agar in the place of gelatin. As a result of her experiences with preserving jams and jellies, she shared her knowledge on the superior qualities of agar compared with gelatin. It required high temperatures for melting, but after cooling down to about 40° C it solidified while still maintaining its structure and relatively transparent form. Agar is a manufactured product derived from various species of seaweed and known in the early 1900s as agar-agar (a Malaysian word meaning "vegetable").

The other individual was another physician working in Koch's lab as an assistant. He modified Koch's glass slides technique by pouring hot agar-based media into glass plates and covered with lids to reduce air-borne contamination. His name? Richard Julius Petri, originator of the "petri" plate, a technique still in use today. ***Now you know the rest of the story.***

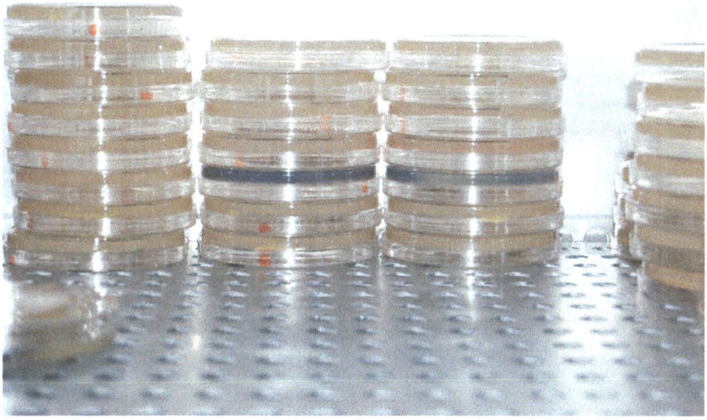

Modern petri plates made of disposable plastic.
Photo Credit: R. M. Harveson

References

Bulloch, W. 1938. The history of bacteriology. Oxford University Press, London, 422 pp.

Hitchens, A. P., and Leikind, M. C. 1939. The introduction of agar-agar into bacteriology. J. Bacteriol. 37: 485-493.

Kado, C. I. 2010. Plant Bacteriology. APS Press, American Phytopathological Society, St. Paul, MN, 336 pp.

Petri, R. J. 1961. A minor modification of the planting technique of Koch. Pages 218-219 in "Milestones in Microbiology". Thomas Brock, ed. Prentice Hall Inc. Englewood Cliffs, N. J.

Smith, E. F. 1905-1914. Bacteria in relation to plant diseases, vol. 1, 285 pp., 1905; vol. 2, 368 pp., 1911, and vol. 3, 309 pp., 1914. Carnegie Institute of Washington.

CHAPTER 2

Bacterial Wilt

One for the Thumb

Introduction

Bacterial wilt of dry beans, caused by *Curtobacterium flaccufaciens* pv. *flaccumfaciens* (*Cff*), has historically posed sporadic but often serious production problems in dry beans throughout the irrigated High Plains since its first report from South Dakota in 1922. In the early 1980s, the disease mysteriously disappeared, appearing only periodically on cull seeds at the processing plants, but with little economic damage.

Suddenly, in late August 2003 before harvest, we found the disease again in two great northern dry bean fields in Scotts Bluff County in western Nebraska. I took dozens of pictures assuming that I would never see it again. However, between 2004 and 2007 the pathogen was additionally identified from hundreds of fields from the central high plains causing severe, measurable yield reductions. Bacterial wilt is now considered an endemic, widely distributed malady in this region and has been documented from well over 500 dry bean fields since 2004.

Recent molecular characterization studies concerning the genetic diversity, population structures and phylogenetic relationships have shown high degrees of heterogeneity among wilt isolates tested. This diversity is perhaps best and most obviously demonstrated by the pathogen's natural tendency to produce different colored colonies growing in culture. The background and history of this unique trait is the rest of the story.

Pathogen Color Variants

Colonies of the original isolates from South Dakota and subsequently other areas of the mid-western U.S. were all yellow in color. However, in 1952, orange-stained seeds were first noted from about 30% of cull samples collected from processing plants after harvest as part of a bacterial disease survey conducted in western Nebraska by Max Schuster (UNL plant pathologist). These discolored seeds subsequently yielded an orange-colored bacterium in culture.

Great northern seeds stained by various pathogen color variants. Reading clockwise from the top: yellow, uninfected, orange, and purple. A pink-stained seed is placed in the box in the upper left corner.
Photo Credit: R. M. Harveson

The orange strains could not be distinguished morphologically from the original yellow strains other than by colony color, and this feature remained stable through multiple transfers and inoculations. Schuster also observed that the tested orange strains were much more virulent than the yellow ones, and the incidence of this variant continued to increase throughout the 1950s and into the 1960s. Schuster interpreted these observations to mean that the differences that did exist between the orange and yellow strains were probably varietal and not specific. Thus he designated this "strain" as *Corynebaterium flaccumfaciens* var. *auranticum* (orange-colored).

It was not until 1967-1968 that another color variant of the pathogen was identified from western Nebraska and reported by Schuster and Anne Vidaver. Infected great northern beans were stained a purple color and were noticed within cull bean samples at processing. Growers and processors remarked that this phenomenon had occasionally been noted since the late 1950s, and the new pathogen was named *C. flaccumfaciens* var. *violaceae* (violet-colored).

The purple variants of *Cff* are actually quite different from the orange and yellow ones. They are less virulent and maintain a yellow color in culture, but they also produce a blue to purple, water-soluble pigment that diffuses into the medium. The purple variant is very rare in incidence, and has only been reported once outside of the western Nebraska Panhandle. Furthermore, the purple pigments from this variant are often unstable and inconsistently expressed in culture. Without the pigment production, these isolates are indistinguishable from other yellow strains, which may explain this variant's lower reported incidence.

Following the 2007 growing season, we recovered a pink bacterial isolate, closely resembling the wilt pathogen, from an orange-stained seed originating from research plots at the Scottsbluff Ag Lab, a research farm for University of Nebraska's Panhandle Research and Extension Center, located northwest of Scottsbluff. The bacterium's identity was confirmed as *Cff* based on morphological, physiological and pathological characteristics. The pink strain was highly virulent on dry beans, producing virtually identical symptoms when compared to a standard known virulent orange isolate after inoculation. This unique trait is very stable, like that of the orange variant, retaining its Pepto-Bismol-colored colonies and virulence after multiple culture transfers.

What is it about western Nebraska's Panhandle?

It is interesting to me to note that 3 of the 4 variants were first identified from western Nebraska, but the orange and purple strains have now been reported from other locations. To my knowledge, the pink isolate has still never been documented from anywhere else, and it coincidentally emerged in the same general area of Nebraska (northwest Scotts Bluff County) where the original orange and purple variants were first found in 1952 and 1967, respectively. Since 2014, additional pink isolates have been recovered from pink-stained dry bean seeds and lesions associated with Goss'wilt-infected corn leaves–once again from field plots on the Scottsbluff Ag Lab research farm.

New Variant

In 2016, a research group in Iran reported the finding of a new red-colored *Cff* isolate, the first and only account to date of this characteristic. Curiously, the red variant was also isolated from an orange-colored seed after harvest, similar to that of the pink variant we first discovered in 2007. The discovery of a fifth color variant, combined with ongoing reports on the highly variable molecular genetics and phylogentic relationships among *Cff* isolates further demonstrate the remarkable diversity of this fascinating bacterial pathogen. ***Now you know the rest of the story.***

Foliar symptoms of severely infected dry bean plants caused by the bacterial wilt pathogen, *Curtobacterium flaccumfaciens* pv. *flaccumfaciens*.*Photo Credit: R. M. Harveson*

References

Agarkova, I.V., Lambrecht, P.A,, Vidaver, A.K., and Harveson, R.M.. 2012. Genetic diversity among *Curtobacterium flaccumfaciens* pv. *flaccumfaciens* populations in the American high plains. Can J Microbiol. 58:788-801.

Harveson, R. M., and Vidaver, A. K. 2008. A new color variant of the dry bean bacterial wilt pathogen (*Curtobacterium flaccumfaciens* pv. *flaccumfaciens*) found in western Nebraska. Online. Plant Health Progress, doi: 10.1094/PHP-2008-0815-01- BR.

Harveson, R. M., Schwartz, H., F., Urrea, C. A., and Yonts, C. D. 2015. Bacterial wilt of dry beans: past, present, and future. Plant Dis. 99: 1665-1677.

Hedges, F. 1922. A bacterial wilt of the bean caused by *Bacterium flaccumfaciens* nov. sp. Science LV: 433-434.

Huang, H. C.. Erickson, R. S., Yanke, L. J., Chelle, C. D. and Mündel, H.-H. 2006. First report of the purple variant of *Curtobacterium flaccumfaciens* pv. *flaccumfaciens*, causal agent of bacterial wilt of bean, in Canada. Phytopathology 90: 1262.

Osdaghi E., Taghavi, S. M., Hamzehzarghani, H., Fazliarab, A., Harveson. R. M., and Lamichhane, J. R. 2016. Occurrence and characterization of a new red-pigmented variant of *Curtobacterium flaccumfaciens*, the causal agent of bacterial wilt of edible dry beans in Iran. Eur. J. Plant Pathol. 146:129–145.

Schuster, M. L., and Christiansen, D. W. 1957. An orange-colored strain of *Corynebacterium flaccumfaciens* causing bean wilt. Phytopathology 47: 51-53.

Schuster, M. L., and Sayre, R. M. 1967. A coryneform bacterium induces purple-colored seed and leaf hypertrophy of *Phaseolus vulgaris* and other Leguminosae. Phytopathol. 57: 1064-1066.

Schuster, M. L., Vidaver, A. K., and Mandel, M. 1968. A purple pigment-producing bean wilt bacterium *Corynebacterium flaccumfaciens* var. *violaceum* n. var. Can. J. Microbiol. 14: 423-427.

CHAPTER 3

The Curious Story of Goss' Wilt of Corn and Bacterial Wilt of Dry Beans

Introduction

Goss' wilt and leaf blight is a destructive bacterial disease of corn, caused by *Clavibacter michiganensis* subsp. *nebraskense*. It was first identified in 1969 from Dawson County near Lexington, NE making it a true native. Over the next decade, it was identified from at least 53 other counties within Nebraska and additionally spread into the surrounding states of Iowa, South Dakota, Kansas, and Colorado. The disease was eventually identified from additional U.S. corn-growing states, before disappearing in the mid-1980s just as suddenly as it first appeared.

Bacterial wilt of dry beans, caused by *Curtobacterium flaccumfaciens* pv. *flaccumfaciens*, has been a serious disease problem. However, its arrivals have been irregular since its initial discovery in 1922 from South Dakota. During the 1940s, its incidences expanded throughout the bean-producing states, most notably in the high plains. The disease persisted into the early 1970s, becoming an endemic and economically important production constraint in Nebraska. This was presumably due to seedborne transmission and dissemination, before disappearing by the early 1980s similarly to Goss' wilt.

Pathogen Re-emergence

However, during the mid-2000s, both pathogens re-emerged to cause serious production problems in Nebraska and other areas of the High Plains wherever corn or dry beans were grown. The resurrection of both diseases simultaneously after a 25-year absence was puzzling. Since that time, these diseases have been investigated extensively by Tamra Jackson-Ziems and myself for Goss' wilt and bean wilt, respectively. Based on these studies

Wilting symptoms on dry bean plants in a field severely infected by bacterial blight. *Photo Credit: R. M. Harveson.*

and empirical observations, we have developed hypotheses as to why these diseases suddenly returned within the same time period, and the details of this narrative are the rest of the story.

Similarities

The two pathogens share several similar characteristics. They are distantly related, and in fact were once classified in the same genus (*Corynebacterium*) primarily due to a superficially similar particle shape, referred to as "coryneform". This characteristic trait gives the genus its name from the Greek noun "koruna" or "club". They both are Gram positive with irregular, short rods that may be bent, slightly curved or club shaped. Disease development for both pathogens, is amplified by injury or stress placed on the plant, and these conditions in Nebraska are often provided by storms consisting of driving rain, hail, wind damage, and sandblasting. Both bacterial pathogens can also become systemic, blocking water translocation within the vascular system of plants, resulting in wilting symptoms; and both favor warm temperatures for disease development after infections occur. Finally, both pathogens can survive in infected residues on the soil surface, serving as the primary inoculum source for later infections. Therefore, infection and spread throughout fields is enhanced by practices such as reduced tillage, continuous cropping, and sprinkler irrigation.

Factors Responsible for Disease Outbreaks?

One of the primary factors, and likely the most influential one responsible for the new epidemics in the mid-2000s was the change in cultural practices adopted in this region 15-20 years previously. Over this time period, the majority of producers introduced some form of reduced tillage into their systems. Combining this practice with the simultaneous, region-wide increases in center pivot irrigation systems would reputably provide better conditions for elevating pathogen survival, infection, and dispersal within fields.

Another factor driving this re-emergence story we can explain conjecturally with the climatic patterns observed throughout the region over this same period. The mid-2000s were characterized by warmer winters, and an extended drought with increasingly higher summer temperatures during the growing season. These factors could have easily contributed to better conditions for pathogen overwintering survival and increased plant stresses, both of which are well recognized to favor disease development and increased severity for both diseases.

Conclusion

Our studies suggest that it is no coincidence that these pathogens reappeared widely throughout this region at roughly the same time. It seems likely that neither disease was completely eradicated, but survived at low levels as saprophytes on weed species or crop residues. Furthermore, we assert that neither disease was routinely noticed until the mid-2000s because formally, the plowing of fields was commonplace, removing a major mechanism for pathogen survival.

During the decade leading up to 2004-2005, reduced tillage became a widespread practice, as was the enormous increase in center pivots in production fields. Since both diseases are more problematic under elevated levels of plant stress, it is also possible that the drought in the mid-2000's played a role. In summary, we have hypothesized that a perfect storm was created with the combination of environmental stress and changes in cultural practices. Cumulatively, all these factors contributed to the high visibility and incidence of both diseases in Nebraska and the Central High Plains in the mid-2000s. ***Now you know the rest of the story.***

Foliar symptoms on corn leaf caused by Goss' wilt. *Photo credit: R. M. Harveson*

References

Harveson, R. M., Vidaver, A. K., and Schwartz, H. F. 2005. Bacterial wilt of dry beans in western Nebraska. NebGuide G05-1562-A.

Harveson, R. M., Schwartz, H. F., Vidaver, A. K., Lambrecht, P. A. and Otto, K. 2006. New outbreaks of bacterial wilt of dry Beans in Nebraska observed from field infections. Plant Dis. 90: 681.

Harveson, R. M., Schwartz, H., F., Urrea, C. A., and Yonts, C. D. 2015. Bacterial wilt of dry beans: past, present, and future. Plant Dis. 99: 1665-1677.

Jackson, T. and Harveson, R. M., and Vidaver, A. K. 2007. Goss's Bacterial Wilt and Leaf Blight of Corn. NebGuide G1675.

CHAPTER 4

The First Disease Forecasting System?

Stewart's wilt is a bacterial disease involving sweet corn as the primary host, but field corn and popcorn can also be affected to lesser extents. All agronomic types of *Zea mays* are susceptible; however different sub-species, hybrids, and inbreds vary in degree of susceptibility.

In the U.S., the disease is endemic throughout the mid-Atlantic and Ohio River Valley, with common appearances in the southern and eastern areas of the Corn Belt, but in general, it has had little effect overall on resistant field corn hybrids in North America. However, under the proper environmental conditions, the disease has historically been seriously damaging on susceptible sweet corn hybrids.

The disease was initially called "bacterial wilt of sweet corn" after its discovery in the mid-1890s. Since that time, it has been called a variety of names, but today it is generally termed "Stewart's bacterial wilt of corn" or simply "Stewart's wilt." Contemporary readers are likely to be familiar with this disease, but did you know that it is also considered by many to be the first disease for which a forecasting system was developed? This account is the rest of the story.

Disease Forecasting

Disease forecasting is a management system that is designed to predict the occurrence and potential severity of plant diseases. These types of systems may be utilized by growers to assist in making economic decisions about treatment for disease control. They are based on educated assumptions involving the pathogen's interactions with the host and environment, with the ultimate objective to accurately predict when the three factors necessary for disease development-host, environment, and pathogen-all interact in such a fashion that disease can occur and cause economic losses.

Discovery of Stewart's Wilt

In 1895, F. C. Stewart with the New York Agricultural Experiment Station first noted a wilt disease causing substantial problems for sweet corn growers in market gardens of Long Island, NY. It was originally identified from an early dwarf variety of sweet corn called 'Manhattan', from Queens, NY causing up to 100% losses. Subsequently, the same disease was found affecting many different varieties abundantly distributed throughout all areas of Long Island.

Stewart accurately described disease symptoms, and fulfilled Koch's postulates by isolating the pathogen in pure culture from diseased plants, and reproducing disease symptoms from inoculated sweet corn plants. A culture was sent to America's first great plant pathologist, E.F. Smith in 1898 who confirmed Stewart's work, and named the pathogen *Pseudomonas stewartii* in Stewart's honor (now known as *Pantoea stewartii* subsp. *stewartii*).

Fred Carleton Stewart
Photo credit: Cornell Mycology Herbarium.

Pathogen Life Cycle

We now are well aware that the pathogen survives within hibernating adult corn flea beetles (*Chaetocnema pulicaria*), residing in the alimentary track before being transmitted into corn plants as a result of beetle feeding. As beetles emerge in spring, they feed on seedlings of early planted corn while also rapidly laying new eggs on the same plants. Both first and second generations of beetles can ingest bacteria as they feed before spreading the pathogen to other plants, and the bacteria can remain in the insect for its entire life.

Stevens' Hypothesis

In the early 1930s, N. E. Stevens noted that a strong relationship existed between winter severity and corn flea beetle survival. He then correlated the mild winter of 1932-1933 with severe bacterial wilt epidemics in 1933 and first published his results in 1934. Over the next several years, he found he could accurately predict the incidence and severity of the disease by simply monitoring the weather conditions during the previous winter. Since the pathogen overwinters and survives within the body of the beetle, he theorized that disease should be less severe following harsh winters, presumably due to fewer numbers of surviving beetles.

His hypothesis empirically predicted that disease would be generally absent if the preceding winter index was below 90°F, and conversely would be present in destructive amounts if the winter index exceeded 100°F. This winter index was calculated from the sum of the mean temperatures occurring during the months of December, January, and February.

Over time this empirically-derived hypothesis proved to be very accurate and beginning in winter of 1934-1935, his predictions were published annually in the Plant Disease Reporter. However, by 1940, this system was discontinued due to a loss of interest for it. Stevens himself stated "it had proven too easy".

Conclusions

Genetic resistance to the Stewart's wilt pathogen incorporated into new hybrids has proven to be the most effective method of disease control, thus the widespread and consistent increase in planting resistant hybrids additionally contributed to the elimination of popularity for Steven's system.

Early efforts at disease forecasting in the 1920s focused primarily on estimating the inoculum potential (amount or concentration of the pathogen) present within fields. As disease control became more sophisticated, another type of forecasting was developed that involved the study and analysis of weather patterns and the pathogen's life cycle. Stewart's wilt of corn is generally considered the first example of developing a forecasting system based on the combination of both environmental modeling and inoculum potential estimation as a pre-plant management method. ***Now you know the rest of the story.***

A young sweet corn plant infected in the seedling wilt phase showing the effects of vascular colonization of the pathogen (wilting and stunting). *Photo credit: R. M. Harveson*

References

Harveson, R. M. 2015. The bacterium of many colors. APS PRESS, St. Paul, MN, 288 pp.

Pataky, J. K. 2003. Stewart's Wilt of Corn. APSnet Features. Online. doi: 10.1094/APSnetFeature-2003-0703.

Pepper, E. H. 1967. Stewart's bacterial wilt of corn. Monograph 4. American Phytopathological Society. St. Paul, MN.

Stevens, N. E. 1934. Stewart's disease in relation to winter temperatures. Plant Dis. Reporter 18:141-149.

Stevens, N. E. and Haenseler, C. M. 1941. Incidence of bacterial wilt of sweet corn. 1935-1940; forecasts and performance. Plant Dis. Reptr. 25: 152-157.

CHAPTER 5

Fire Blight

The First Phytobacterial Disease?

Introduction and Background

It is likely that somewhere in our education and training as plant pathologists, we have all been told that Thomas Jonathan Burrill, a professor of botany at the University of Illinois, was the investigator that furnished the first credible proof that plant diseases could be caused by bacteria. That concept is not completely accurate, and in order to understand the whole picture, you must hear the rest of the story.

By the 1870s, plant pathology studies centered on the fungi. The veracity of Anton de Bary's studies with potato late blight from the 1850's was now universally recognized and accepted, proving that fungi could be plant pathogens. Many of the severely damaging diseases of economic importance in crops were caused by fungi that were easily observed–the rusts, smuts, and mildews.

Thomas J. Burrill
Courtesy of Wikipedia

Therefore, fungi were easy targets for placing the blame for almost any plant disease. Furthermore, the majority of the plant scientific community, centered in Europe believed that bacteria were not primary causal agents of plant disease. Instead, they insisted that bacteria were merely contaminants and secondary colonists of plant tissues infected by fungi or injured by other causes.

Robert Koch and his Postulates

In the mid-1870s, the German physician, Robert Koch demonstrated that the anthrax disease in cattle was caused by bacteria. Previous researchers had compared diseased animals and noted the presence of the anthrax bacterium in their blood. However, Koch was the first to experimentally prove the cause and effect of this relationship. Koch's major influence on plant bacteriology was twofold: 1) improvement of pure culture methodologies; 2) design of steps necessary to determine the cause of a disease, now known as Koch's postulates.

Fire Blight

Fire blight is a disease of rosaceous fruit crops, primarily pear, apple, and quince, caused by the bacterial pathogen *Erwinia amylovora*. It has been historically tied to the United States and is furthermore assumed to be native to North America since it has been observed and recorded in the U.S. since colonial times. It became problematic to commercial apple and pear production in the 1820s and spread to the west coast of the U.S. by the 1880s.

Burrill began working informally on fire blight in 1873 and the results from his studies were presented largely in papers and discussions with the Illinois Horticultural Society in the years 1878-1884. He conducted extensive disease transmission experiments successfully but at least 3 different workers have been documented as transmitting fire blight earlier–E. S. Hull in 1870, R. Ragan in 1846, and H. Wendell in1850.

Other Phytobacterial Investigators

A number of other contemporary workers were also investigating phytobacterial diseases during this time. F. M. Dränert studied a bacterial disease of sugar cane in Brazil (1869) that is now known as *Xanthomonas axonopodis* pv. *vasculorum*. He did not produce pure cultures or attempt inoculation, but is still credited with observing bacterial cells within diseased tissues. Edouard E. Prillieux, a professor at the National Institute of Agronomy in Paris, published a report on the rose-red or micrococcus disease in 1879. It consisted of a microscopic study on diseased wheat kernels, where he observed masses of micrococcus-like organisms forming distinct cavities within seeds. Orazio Comes was teacher of botany at the Royal Agricultural College in Portici, Italy. He published several papers on what he called *Bacterium gummis*, which was widely distributed on numerous crops. Neither Prillieux nor Comes was able to grow their respective organisms in pure culture, achieve disease transmission, or adequately describe the pathogens, so today we are uncertain of the identity of those diseases on which they were working.

Texas pear orchard affected by fire blight (1902). *Courtesy M. B. Waite, U. S. Department of Agriculture*

As early as 1882, Jan H. Wakker in Holland identified a yellow bacterium from infected hyacinth bulbs, named it *Bacillus hyacinthi* (later re-named *Xanthomonas*) and published his results in 1883, followed by four additional research papers on this disease between 1884 and 1887. They were not well recognized then or now as two were published in German, three in Dutch, and one in French. Wakker was the first to try to prove that bacteria-caused plant disease by the new culture methods, first publishing his results a mere seven years after Koch proved that bacteria caused disease in animals, but he did not make successful inoculations with pure cultures until 1886. In 1887, another Italian, Luigi Savastano conclusively demonstrated the constant association of bacteria in the stem knots, isolated them in pure culture, and then produced typical galls after introducing into small quantities of the culture into healthy olive shoots. Both Wakker and Savastano independently produced results superior to those of Prilleaux, Comes, and Burrill, but still trailed in terms of reporting.

Contributions of Arthur

By 1883, Burrill was convinced that the fire blight disease was caused by a bacterium, for which he named *Micrococcus amylovorous*. He also demonstrated the association of the bacterium with diseases on pear, apple, and quince, but inexplicably, never used pure cultures, nor attempted to employ Koch's postulates for proof of pathogenicity as would be required today. Thus, it was left for J. C. Arthur to complete the link between bacteria and the fire blight disease.

In 1884, Joseph Charles Arthur, as the botanist of the New York experiment station at Geneva, repeated Burrill's fire blight experiments with more extensive and detailed studies and reported them in 1885. Using the newly developed plate culture techniques, Arthur cultured the bacterium in a sterile infusion of corn meal and first infected green pear fruit, and later, healthy trees. He also filtered a suspension of the blighted pear through a porous earthenware jar and obtained infection after inoculation with the unfiltered liquid but not from the filtrate. In 1886, he submitted the results of his fire blight studies to Cornell University as a doctoral thesis before moving on to Purdue University, where he eventually became famous as one of America's greatest rust investigators. Arthur's convincing evidence settled the point with pure culture methods, but he provided no positive evidence that the pathogen was re-isolated from inoculated plants (final step of Koch's postulates).

Burrill's Legacy

Although unquestionably groundbreaking, Burrill's results for proving bacterial pathogenicity in plants were not backed by today's more exacting standards. But why is he still so universally praised and credited with this honor? The eminent American phytobacteriologist, E. F. Smith stated that "just as Pasteur's contribution to science is more vital than Koch's because it was earlier and was pioneer work, so Burrill's discovery was more difficult to make and hence more worthy of praise than anything that has come after". It has been postulated that Smith's reputation and influence in the scientific community was so great that his evaluations and statements about Burrill were uncritically accepted by successive instructors and textbook authors.

Discovery is often determined by time, fortune, and fortuitous circumstances; and advancements are often built upon the work of previous predecessors. Nevertheless, if any one person must be singled out as having proved that bacteria do cause disease in plants, it would likely have to be Arthur rather than Burrill. However, many investigators between 1846 and 1901, now largely forgotten by history, also made vital contributions toward the elucidation and acceptance of the concept. ***Now you know the rest of the story.***

References

Ainsworth, G. C. 1981. Introduction to the history of plant pathology. Cambridge University Press, Cambridge.

Baker, K. F. 1971. Fire blight of pome fruits: the genesis of the concept that bacteria can be pathogenic to plants. Hilgardia 40: 603-633.

Griffith, C. S., Sutton, T. B., and Peterson, P. D. 2003. Fire blight. The foundation of phytobacteriology. APS Press, American Phytopathological Society, St. Paul, MN, 144 pp.

Kennedy, B. W., Widin, D. D., and Baker, I. F. 1979. Bacteria and the cause of diseases in plants: a historical perspective.

Smith, E. F. 1896. The bacterial diseases of plants: a critical review of the present state of our knowledge. Am. Nat. 30: 626-643

CHAPTER 6

Those Villainous Vexations of the Vine

The use of fungicides to protect plants from fungal pathogens developed slowly and empirically over hundreds of years. According to James Horsfall, the earliest reference to a fungicide was the utilization of amurca of olives, made by Pliny in 470 BC for the prevention of blight infection (presumably powdery mildew).

After the germ theory was established in the 1850s, the development and use of fungicides substantially accelerated. This was partially due to the improved knowledge concerning the principles of plant pathology and disease. However, its primary driving force was in response to the accidental introduction of several North American grape diseases and pests into Europe. These epidemics seriously threatened the survival of the important European wine industry and are the rest of the story.

The Three Great Grape Epidemics

Three major epidemics occurred in European vineyards in the latter half of the 19th century, which cumulatively threatened to end viniculture, particularly in France. All three problems are thought to have originated in North America on wild species of *Vitus*, prior to emerging in Europe. The severity of damage caused by these issues greatly influenced the development and improvement of fungicides for controlling plant diseases.

Powdery Mildew

The first great epidemic was powdery mildew, and is assumed to have been introduced into Britain from North America on vines for experimentation. The pathogen's (*Erysiphe necator*) origin is uncertain but is known to have existed on wild species of grapes in

Powdery mildew on a grape leaf *Photo Credit R. M. Harveson*

the eastern U.S. for centuries. The most susceptible species were presumably killed off, but their surviving descendants developed a strong resistance to the disease.

In 1845, an English gardener named Edward Tucker noted a powdery substance on wine grapes being raised in a greenhouse resembling similar symptoms seen on peach trees several years before. Applications of sulfur and lime solution successfully halted progression of both diseases.

In 1848, the disease was first identified in France and subsequently threatened the vineyards of the entire continent. Tucker's lime-sulfur treatments worked well for small numbers of plants in the greenhouses. However, it was not practical on a field scale involving millions of acres of grapes, requiring spraying each new flush of foliage up to 3 times during the season. The only viable option for field treatment at that time was applications of sulfur dust to the infected vines, which kept the disease reasonably in check.

Phylloxera

As a result of the powdery mildew epidemic, native American grapes, largely immune, but agronomically unsuitable, were imported for hybridization with high quality European wine varieties. This practice then introduced the second major epidemic from the Americas suffered by European grape production, known as "Phylloxera" (*Daktulosphaira vitifoliae*, family Phylloxweridae). Phylloxera is an almost microscopic, pale-yellow sap-sucking insect (related to aphids) that feeds on the roots and leaves of grapevines. Their feeding slowly kills plants by cutting off water and nutrient flow in the vascular system of vines.

Chemical treatments attempted for control failed, as did other bizarre remedies, including human urine, arsenic compounds, tar, and even high voltage electricity. The issue became so severe that the French government actually offered a never claimed reward (300,000 francs) for anyone who could solve the problem.

This complication was eventually solved by producing new hybrid grape varieties combining European scions grafted onto resistant American root stocks. Fortunately, the advance of the Phylloxera was slow enough to allow the time-consuming creation and replanting of new vines to proceed. Although this process saved the industry, this insect pest is still present in every European vineyard today.

Downy Mildew

The third and last epidemic of the French vineyards originating from North America was, in turn, inadvertently introduced in about 1878 with the thousands of Phylloxera-resistant grape root stocks brought to address the root insect infestation. Downy mildew, caused by the oomycete *Plasmopara viticola*, was an endemic resident of the wild *Vitis* species in North America, but after introduction into Europe it became the most serious and destructive of the three epidemics from America.

Due to the cool and wet conditions during the early 1880s

the downy mildew pathogen flourished and spread very rapidly throughout France. The newly generated plants were resistant to Phylloxera, but were particularly susceptible to downy mildew. Without the eventual inauguration of the Bordeaux mixture in French vineyards, economic losses would likely have been catastrophic. Unlike powdery mildew, downy mildew was not effectively managed by sulfur alone, which became the driving force for experimentation with alternative fungicides. ***Now you know the rest of the story.***

Downy mildew symptoms showing yellow lesions on upper leaf surface (left) with the corresponding lower surface consisting of the whitish "downy" sporangia emerging from stomata (right). *Photo credit: USDA Yearbook of Agriculture, 1953.*

References

Ainsworth, G.C. 1981. Introduction to the history of plant pathology. Cambridge University Press, Cambridge.

Horsfall, J.G. 1945. Fungicides and their action. Chronica Botanica Company, Waltham MA, 239 pp.

Large, E.C. 2003. The advance of the fungi. APS Press, St. Paul, MN, 488 pp.

Schuman, G.L. 1991. Plant diseases: their biology and social impact. APS Press, St. Paul, MN, 397 pp.

CHAPTER 7

Rhizoctonia spp.

The First Documented Root Pathogen

The genus *Rhizoctonia* is comprised of a group of highly variable, soilborne, sterile fungi, yet they still share a number of similar characteristics. Most notably, they are long-lived in soils, remaining in a vegetative state, rarely making any reproductive spores. Species of *Rhizoctonia* are highly effective saprophytes, while also being economically important root pathogens on many crops. In fact, the ability for this fungus to aggressively attack a wide range of host roots is responsible for its generic name, *Rhizoctonia*, which is derived from Greek, translating into English as "killer of roots".

Rhizoctonia solani

Rhizoctonia solani is the most well-known species, and thought to be present throughout most if not all of the world's cultivable landmass. It causes a myriad of diseases, including seed decay, post- and pre-emergent damping-off, stem cankers, root and crown rots, fruit decay, and in some instances, foliar diseases.

Sugar beet infected by *R. solani*, causal agent of Rhizoctonia root rot. *Photo credit: R. M. Harveson*

The history of *R. solani* is almost as long as that of modern plant pathology itself, being first implicated as a root pathogen within the same general time period of the Irish potato famine (1850s). However, it was not the first *Rhizoctonia* pathogen discovered. You may be surprised to learn that a lesser-known species of *Rhizoctonia* had been noted causing a root disease called violet root rot 130 years earlier. That is the rest of the story.

Rhizoctonia crocorum

In 1728, Henri-Louis Du Hammel de Monceau described a disease of the crocus plant, *Crocus sativa*, in southern France, that proved to be the first report for an ailment due to *Rhizoctonia*. This peculiar plant is a member of the iris family, and the source of the spice saffron. Diseased bulbs were shrouded with a thick reddish-violet hyphal mat with knot-like swellings (sclerotia). He also noted that the fungus moved spatially through the soil, infecting, and spreading disease from bulb to bulb underground. Fifty years later, another Frenchman named Fougeroux de Bondaray reported a disease appearing on asparagus after being grown on land previously cultivated with similarly infected crocus bulbs.

The French mycologist Augustin Pyramus De Candolle formally created the genus *Rhizoctonia* in 1815 to accommodate the crocus pathogen (providing the species name), naming it *R. crocorum.* De Candolle additionally recognized a similar disease of alfalfa producing a mat of violet mycelium on roots of affected plants, naming it *R. medicaginis.* The Tulasne brothers (Charles and Louis Rene) considered the crocus and alfalfa pathogens to be identical, and reduced the two species to a single, more descriptive name, *R. violacea* in 1851. Nevertheless, the designation, *R. crocorum*, remained since it had priority over the other species.

Violet Root Rot in Europe

Julius Kuhn published an account in 1858 of the appearance of a Rhizoctonia disease in Germany causing a serious root rot on alfalfa, carrots, and mangolds (early sugar beets) accompanied by a purple mold on subterranean plant parts. It was distributed in localized clusters within fields, yet still caused substantial wilting and rapid death of plants in large areas. This condition was referred to as "Fehlstellen", translated into English as "dropouts." The disease became well known throughout Europe by various names like violet root rot, red root rot, or violet root felt disease, afflicting at least 47 genera of plants in 21 families (mostly dicots).

Violet root rot in the U.S.

In the U.S., just as in Europe, the first Rhizoctonia disease discovered was violet root rot. This dubious distinction was attributed to Nebraska from alfalfa. Curiously, reports were made independently in 1890 from two sites by two different individuals (Roscoe Pound and W. J. Webber) in Nebraska. Both referred to the causal agent as *R. medicaginis* additionally remarking that it was rare, of little consequence and "seemed to be harmless".

Sugar beet infected with *R. crocorum,* causal agent of violet root rot. *Photo credit: R. M. Harveson*

Conversely, G. F. Freeman reported in 1908 that the red root disease of alfalfa was spreading rapidly in the state, potentially threatening the industry in Kansas, and every effort should be made to find a practical means of control. In 1915 it was first reported on potatoes in Oregon, and was additionally found infecting potatoes throughout Western Nebraska between 1916 and 1920, by George Peltier, where it was described as "very prevalent".

Violet Root Rot Today

Since that time, the disease has been found sporadically from across the U.S. on potatoes, carrots, sugar beets, alfalfa, and celery, but is still considered to be relatively rare, unlike the pervasively occurring *R. solani.* It is also interesting to note how the two *Rhizoctonia* species differed in their incidence and distribution. Despite being the first species identified from both locations, *R. crocorum* was predominant in Europe while *R. solani* was more habitual in America. ***Now you know the rest of the story.***

References

Duggar, B. M. 1915. *Rhizoctonia crocorum* (Pers.) DC and *R. solani* Kuhn (Corticum vagum B. & C.) with notes on other species. Ann. Missouri Bot. Gardens 2: 403-458.

Harveson, R. M. 2015. The bacterium of many colors. APS Press, St Paul, MN, 288 pp.

Peltier, G. L. 1916. Parasitic Rhizoctonias in America. Illinois Agric. Exp. Sta. Bull. 189: 283-390.

Zadoks, J. C. 1981. Mr. Duhamel's 1728 treatise on the violet root rot of saffron crocus: "Physical explanation of a disease that perishes several plants in the Gastinois [France], and saffron in particular". Mededelingen van de Landbouwhogeschool, Wageningen 81-7: 1-31.

CHAPTER 8

Dry Rot Canker

An Obscure Rhizoctonia Disease Little-Known for a Century

In another chapter in this book on *Rhizoctonia* as the first root rot pathogen identified, I related to you a brief synopsis of the rarely occurring root disease (at least in North America), violet root rot, its causal agent, *Rhizoctonia crocorum*, and its relationship with *R. solani* as pathogens. This chapter will focus on an even more uncommon Rhizoctonia disease of sugar beets called dry rot canker (DRC) for which I have personal experience. It is virtually unknown, even to veterans of the sugar beet production industry. The cause of DRC is now known to be a largely uncharacterized species of *Rhizoctonia*, referred to as "binucleate" and its history, past and recent, is the rest of the story.

Symptoms of dry rot canker on a sugar beet root. Note the internal dried rot within root. *Photo credit: R. M. Harveson*

New Disease

B. L. Richards first reported a previously undescribed root disease of sugar beet near Cornish, Utah in August 1920 calling it dry rot canker. The disease was characterized by localized, dry sunken lesions that penetrated deeply into the taproot interior, causing the decaying tissue to rapidly dry out as infection continued inward. This resulted in cavities filled with a dry pithy material consisting of both fungal hyphae and decayed host matter. Surface tissues of the cankers also produced a distinctive series of concentric circles, like a target board. These signs and symptoms were quite distinct from those of the well-known *Rhizoctonia* root and crown rot disease (RRCR) caused by *R. solani*.

Richards further mentions that he learned through correspondence with G. L. Peltier that a similar disease was also observed in Nebraska in 1920. Furthermore, based on preserved herbarium specimens, the disease's presence could be traced to Utah as early as 1915. Another similar sugar beet root disease was later identified in Minnesota and Colorado during the summers of 1936-1938. DRC has since occurred infrequently throughout irrigated production areas of the western U.S, including California, Montana, North Dakota, and Wyoming; and until recently, little else was known about this disease, primarily due to its rare appearances.

New Appearances

In mid-September 2011 a field near Bridgeport, in Morrill County, NE, was brought to my attention with a high incidence of plants (10-15%) exhibiting wilting and yellowing symptoms suggestive of RRCR. However, root symptoms were inconsistent with that disease. Symptoms consisted of sunken lesions with concentric circles and a rot penetrating deep into root tissues like those described by Richards as DRC.

Between 2013 and 2015, more than a dozen additional sugar beet fields in Morrill and Scottsbluff Counties of western Nebraska were identified with plants displaying identical symptoms. Isolations from diseased root tissues all yielded fungal cultures strongly resembling *Rhizoctonia*. Koch's postulates were fulfilled in the greenhouse and compared to inoculations with RRCR.

Symptoms of dry rot canker showing the circular, target-like rings. *Photo credit: R. M. Harveson*

Further Studies

Molecular analyses showed that sequences from the ribosomal DNA ITS region of the DRC isolates showed significant homology (96% identity) with sequences of a binucleate *Rhizoctonia* species (AG-Fb). Microscopic examination of these isolates revealed the presence of two nuclei within hyphal cells. All subsequent DRC isolates examined to date (>20), possess near identical DNA sequences.

Based on the different symptoms induced in both the field and greenhouse, nuclear condition, and DNA sequence similarities with binucleate *Rhizoctonia* species, the DRC isolates were proven to be distinct from the multinucleate *R. solani* pathogen associated RRCR. Recent field studies in Nebraska have further demonstrated that the DRC pathogen is sensitive to current fungicides and disease tolerant cultivars in the same manner as *R. solani,* but caused significantly greater yield reductions and disease severity in *Rhizoctonia*-susceptible cultivars compared with those induced by *R. solani.*

Concluding Remarks

Richards, the original investigator, presumed that the *Rhizoctonia* isolates inducing the dry rot canker disease were different than typical *R. solani* isolates based on different symptoms. This remarkably prescient observation was made long before the availability of our current more sensitive molecular tools. Our studies confirmed those suspicions, proving that the pathogen is a distinct *Rhizoctonia* species, while also documenting the emergence of DRC for the first time in 90 years.

Because *R. solani* and other *Rhizoctonia* species do not produce conidia and only rarely produce basidiospores, the classification of these fungi often has been difficult because of a poorly defined taxonomy. It is becoming apparent that this fungal genus is actually a large complex composed of genetically distinct groups with divergent life histories. The DRC pathogen is a good example of one of these largely unexplored variant strains of *Rhizoctonia.* ***Now you know the rest of the story.***

References

Harveson, R. M. 2015. The bacterium of many colors. APS Press, St Paul, MN, 288 pp.

Harveson, R. M., and Bolton, M. B. 2013. First evidence of a binucleate *Rhizoctonia* as the causal agent of dry rot canker of sugar beet in Nebraska. Plant Dis. 97: 1508.

LeClerg, E. L. 1939. Studies on dry-rot canker of sugar beets. Phytopathology 29: 793-800.

Richards, B. L. 1921. A dry rot canker of sugar beets. J. Agr. Res. 22: 47-52.

CHAPTER 9

Discovering the Function of Pycnia in Rust Fungi

Stem rust of wheat, caused by the fungal pathogen *Puccinia graminis f.* sp. *tritici*, is perhaps the most famous (infamous) plant disease of antiquity. Historically, it has caused endless misery to humans, and its effects have been recognized for millennia, although the cause was unknown until the mid-19th century.

The pathogen's life cycle is very complex, involving 5 separate spore stages-pycnial, aecial, uredinial, telial, and basidial), each having their own distinct function. Furthermore, two different plant hosts are required to complete this disease cycle.

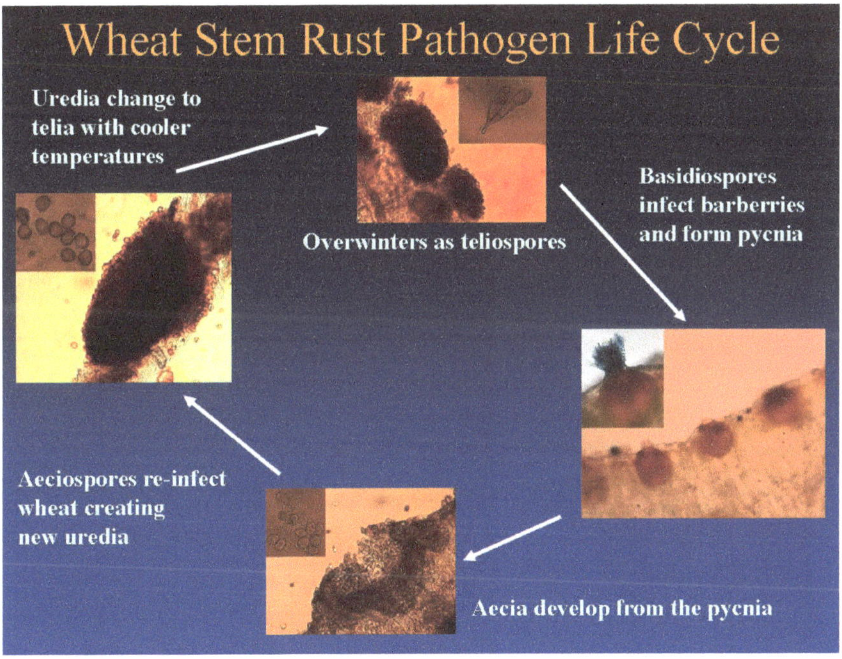

Life cycle of wheat stem rust pathogen. *Photo Credit. R. M. Harveson*

Epidemiological Advancements

The mid-1800s saw a dramatic increase in knowledge concerning the life history of the rust fungi. In 1854, the Tulansne brothers (Charles and Louis René) showed that the uredia and telia (stages found on primary wheat host) did not originate from two separate rust species, as was thought at the time, but were derived from the mycelium of the identical fungus, *P. graminis*.

Farmers first noted the relationship between disease and barberry bushes near wheat fields in the 1600s. However, it was not until 1865 when Anton de Bary formally established that the stem rust pathogen generated those spore stages appearing in a regular sequence and that it required two different, unrelated (alternate) hosts to complete its life cycle, a condition referred to as heteroecious. The alternate host for stem rust is several species of barberry (*Berberis spp.*).

Thin section of a rust-infected leaf showing two adjacent pycnia in the epidermal surface. Note the pycniospores (right) and the receptive hyphae (left) *Photo Credit: R. M. Harveson*

Life Cycle

The fact that the stem rust pathogen spends part of its life on the barberry plant allows the sexual phase of the fungus to occur. This process accommodates development of new strains after genetically different but compatible mating-type nuclei fuse (heterothallism), resulting in hybrid-like combinations of potentially new variety-specific pathogenicity genes. The alternate host (barberry) becomes infected when overwintering teliospores, derived from rust-colored urediniospores from the wheat host germinate and form basidia in the presence of free moisture. Resulting basidiospores can then be transported short distances to infect alternate hosts. Pycnia with pycniospores and receptive hyphae then appear about a week later embedded within orange pycnial lesions on barberry leaves. These spores serve as gametes and fertilization occurs when the nectar exuding from pycnia containing pycniospores is moved to receptive hyphae of the opposite mating type by insects or splashing water. Aecia and aeciospores are formed that re-infect wheat plants, forming new uredinia and completing the cycle.

Much of what we know today about the biology of the rust fungi was first learned from studying stem rust, but the purpose of the pycnial stage was unknown. It was not until 1927 that the functional role of the pycnial stage of the rust pathogen was deduced by John H. Craigie, a graduate student with the Dominion Rust Research Laboratory in Winnipeg, Canada. Many of you may know this story, but did you know that Craigie's milestone discovery did not occur with wheat stem rust, but another similar disease–sunflower rust? That fact is the rest of the story.

Craigie's Landmark Discovery

Craigie's benchmark concept was conceived after watching a fly landing and fluttering about on nectar-filled pycnial pustules of rust-infected sunflower seedlings in the greenhouse (all stages of sunflower rust occur on the sunflower, a condition known as autoecious). He mixed the exudate (nectar) emerging from the pycnia from several pustules, as the fly was assumed to have done. He found that isolated infections arising from single pycniospores produced new pycnia giving rise to pycniospores and nectar, but no aecia developed as they often did when several pycnial pustules were close to one another on leaves.

Craigie concluded that the spores must be different types that he referred to as (+) and (-) and when adjacent infections of single spores were of the opposite type, hyphal fusions then must occur, followed by the development of new aecia. He confirmed this supposition and experimentally demonstrated that aecial formation occurred after transferring spores from one pycnium to another of the opposite mating type, thus elucidating the function of the pycnia and demonstrating heterothallism in the rusts.

Conclusions

John Craigie's crucial breakthrough that identified the function of the pycnial spore stage proved that rust fungi had the ability to reproduce sexually, potentially resulting in new strains (races) of the fungi each season. This explained the baffling problem previously encountered when resistant cultivars yielded well one year, then suffered huge losses due to disease the next. Interestingly, this revelation occurred first from sunflower rust and not wheat stem rust.

It is ironic that this major watershed finding was first made on a plant disease that was not economically important, nor involving a cash crop being grown at that time. Commercial sunflower production in North America would not begin for another 25-30 years. ***Now you know the rest of the story.***

References

Craigie, J. H. 1927. Discovery of the function of the pycnia of the rust fungi. Nature 120: 765-767.

Craigie, J. H. 1927. Experiments in sex on rust fungi. Nature 120: 116-117.

Estey, R. H. 1994. Essays on the early history of plant pathology and mycology in Canada. McGill-Queen's University Press, Quebec, 384 pp.

CHAPTER 10

Puccinia subnitens

An Obscure but Fascinating Rust Pathogen

Introduction

Sugar beet seedling rust, caused by the fungal pathogen, *Puccinia subnitens*, is a disease of Chenopodiaceous plants that has rarely been observed in commercial crop production fields. Due to the pathogen's unique life cycle, it is essentially harmless for the sugar beet industry, despite its name. However, it has caused serious but sporadic losses to spinach growers in Colorado and the Pacific Northwest. The tale of this highly uncommon fungal pathogen is the rest of the story.

Pathogen Life Cycle

Puccinia subnitens is a macrocyclic (produces 5 distinct spore stages) rust fungus that completes its exceedingly complex life cycle on completely different and unrelated hosts (heteroecious). Only two stages (pycnial and aecial) occur on the hosts of economic importance (primarily sugar beets and spinach), while the uredial and telial stages infect feral species of saltgrass (*Distichlis spp.*). Saltgrass is a warm season grass native to arid areas of the western United States, and is also commonly found in brackish, marshy areas or high saline soils.

In early spring, overwintering teliospores germinate from telial pustules on saltgrass to produce basidiospores, which then are blown by wind where they can infect sugar beets or a number of other alternate hosts, including common lambsquarters. Infections by the basidiospores give rise sequentially to the pycnial and aecial spore stages. Flask-shaped pycnia are embedded within circular, light-yellow lesions, generally on the upper leaf surface. After fertilization occurs, the aecia are formed on the lower leaf surface directly below the pycnia. The aecia release spores which re-infect the saltgrass, resulting in new uredia and telia, completing the life cycle.

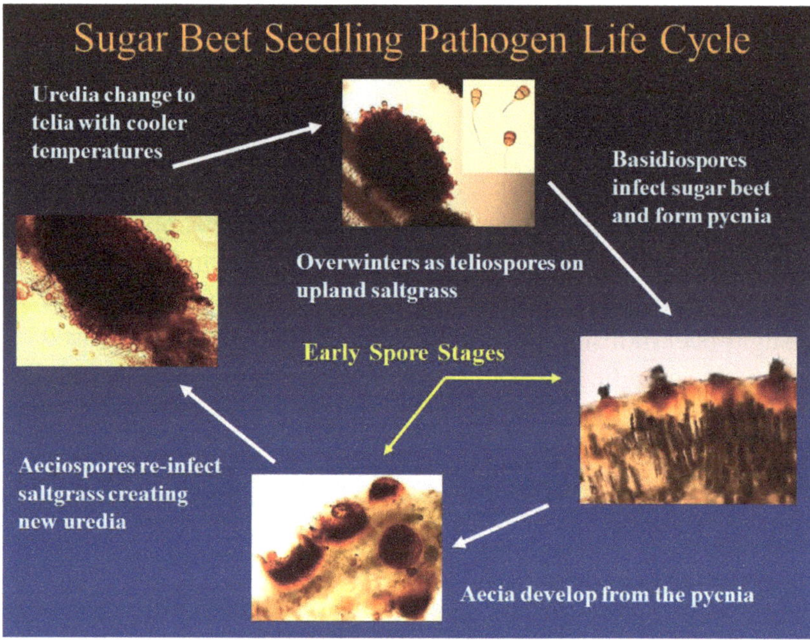

Life cycle of the pathogen causing sugar beet seedling rust.
Photo credit: R. M. Harveson

Reports of the Disease

The first description of *Puccinia subnitens* (telial stage) was by German mycologist Paul Dietel from Montana in 1895, but it was almost a decade before its telial form was correlated with its aecial stage in different plants.

The first published report of this disease on an economic crop (sugar beets) was from Colorado in 1914 by Venus W. Pool and M.

B. McKay after seeing it during the 1912-1913 seasons from multiple fields near Rocky Ford, CO. They also noted lambsquarters populations that were heavily infected with aecial lesions within the same fields. By early July the disease intensity on the sugar beet host had diminished and was no longer an issue. They further added that the telial form was found abundantly on saltgrass stands near sugar beet fields during the epidemics.

In mid-May 2009, young sugar beet and lambsquarters plants were first observed exhibiting signs indicative of this same rust disease in a production field near Bayard, Nebraska after an extended period of unusually cool and wet weather in April and early May. Disease incidence in this field approached 25% although lesions were restricted primarily to the cotyledons, and this field was similarly surrounded by stands of rust-infected saltgrass like the report of Pool and McKay almost 100 years previously.

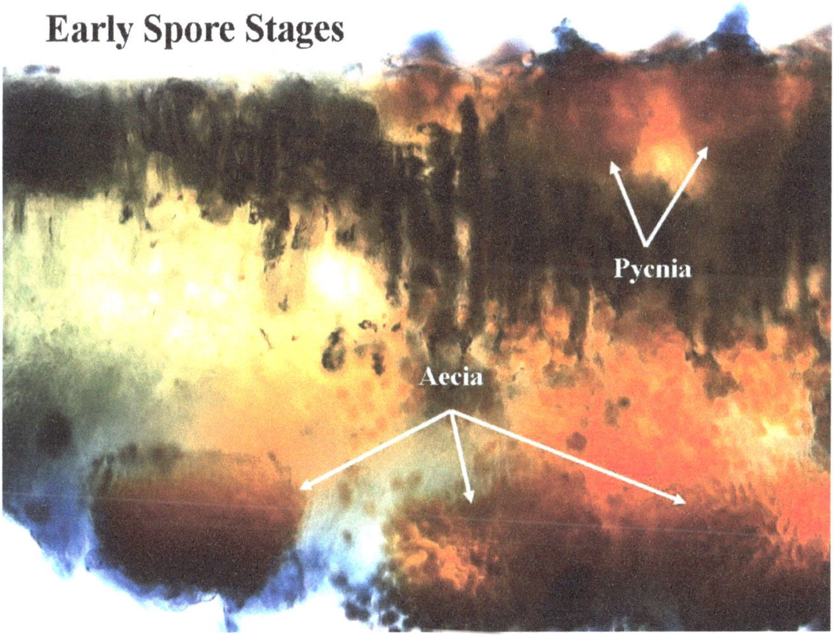

A thin section of a rust-infected sugar beet leaf. Notice the bottle-shaped pycnia on the top of the leaf and the aecial cups producing the aeciospores on the bottom. *Photo credit: R. M. Harveson*

Rust in Spinach

Today we are aware that this disease is not a serious economic restraint to sugar beet production, as it is found early in the season, generally limited to the cotyledons, hence the name "seedling rust". Conversely, it has been documented periodically (1922, 1943-44, 1952, 1959, 1996, 1999) causing severe losses for spinach growers, in part because the aecial-infected leaves are the part of the plant that is directly marketed and consumed at the time of infection. Additionally, infection in spinach late in the season prohibits the crop from "outgrowing" the disease with new foliar growth, as would occur with sugar beets.

Economic damage was also suffered due to the unique exercise of shipping unknowlingly infected spinach plants within cool and moist, enclosed containers until reaching their final market destination. These conditions enabled the disease to progress rapidly, rotting the leaves while in transit, making the product unmarketable upon arrival.

Field-infected, young sugar beet plant exhibiting both pycnial and aecial lesions, but note its restriction to the cotyledons. *Photo credit: R. M. Harveson*

Very Unique Rust Pathogen

Not only is this pathogen rare in appearance, it also causes a distinctly atypical rust disease. Seedling rust is not economically damaging on beets, unlike with spinach, due to botanical and biological differences between the two crops. Infection of the sugar beet occurs very early in the plant's life. Repeated infections from the early spore stages (pycnial and aecial) to newly developing leaves is not possible because of the limitations of the pathogen's life cycle. The aeciospores that form in response to pycnial infection on sugar beets will not re-infect that same crop again that season. They will only infect the alternate host, *Distichlis spp*, while new beet infections that season require another flush of basidiospores from teliospores derived from the alternate hosts, infected the previous season.

Furthermore, this pathogen is also unusual because it is one of the few documented heteroecious rust diseases where the early spore stages (pycnial and aecial) occur on the economic host while the normally damaging, repeating, uredial stage is found on the wild, non-economic host. Most significantly, it was the first heteroecious rust disease recognized to have the ability to infect numerous, distinct host species with the aecial stage while maintaining a relatively narrow host range for its uredial and telial stages.

Heteroecious rusts tend to limit their aecial hosts to species of one genus with the occasional one that may extend to several genera within a single family, but *P. subnitens* has been the notable exception. It is quite remarkable that the now-known host range of *P. subnitens* includes at least 84 plant species in 19 families and 52 genera. ***Now you know the rest of the story.***

References

Barss, H. P. 1922. Destructive rust (*Puccinia subnitens* Dietel.) on spinach in the northwest. Phytopathology 12: 446.

Dietel, P. 1895. New North American Uredineae. Erythea 3: 81: 57-82.

Harveson, R. M. 2010. First Report of Sugar Beet Seedling Rust Caused by *Puccinia subnitens* in Nebraska. Online. Plant Health Progress. doi:10.1094/PHP-2010-0315-03-BR.

Harveson, R. M. 2015. The bacterium of many colors. APS Press, St Paul, MN, 288 pp.

Pool, V. W., and McKay, M. B. 1914. *Puccinia subnitens* on the sugar beet. Phytopathology 4: 204-206.

Ramsey, G. B., Smith, M. A., and Wright, W. R. 1952. Orange rust, a serious market disease of Colorado spinach in 1952.

Skoglund, L. G., and Brown, W. M. Red rust of spinach! 1999. Pest Alert, Vol. 16, No. 8. Cooperative Extension: Colorado State University, Ft. Collins, CO.

CHAPTER 11

Black Rot of Grapes

Plant Pathology Begins in the U.S.

I have several chapters in this book on the Bordeaux mixture and its adoption and usage in Europe for downy mildew of grapes. By the 1880s it had saved vineyards and thus the critically important wine industry throughout the continent. This was particularly true in France, where it became the benchmark of fungicides for many years to come. But did you know that in the United States, this fungicide was originally launched and evaluated as a result of a totally different grape disease called black rot? The study of this disease and efforts to manage it were also instrumental in initiating plant pathology in the U.S. as well as shaping and developing the USDA, which is the rest of this story.

Frank Lamson-Scribner

The USDA was established in 1862 by President Abraham Lincoln, however over the next two decades, work of the agency was largely restricted to publishing informational bulletins, distributing seeds, and assembling agricultural statistics, rather than conducting scientific research. An undercurrent of doubt arose among members of the agricultural media and the public concerning the usefulness and/or even the necessity for keeping the new Federal agency. This notion began to change in 1886 when Frank Lamson-Scribner was appointed as the head of the new Section of Mycology within the USDA, making him the first scientist employed by the Federal Government to work specifically in the burgeoning new discipline of plant pathology.

Scribner believed that science was the fundamental mechanism for achieving ultimate disease control. Therefore, he wanted to transfer the responsibility of proving this hypothesis away from the previous amateur hobbyists to the trained scientists. His approach involved first investigating pathogen life histories and their relationships with the diseases they caused. He also recognized the critical need to develop a template that combined fundamental science with practical application, for the purpose of creating a more appealing political relations victory. Thus, he needed a model with both high visibility and a high probability of success in order to adequately "sell" to the public his concept of this new agency, demonstrate the value of research, and how it would benefit all taxpaying citizens.

Black rot symptoms on grapes. *Photo credit: Plant Diseases, USDA Year Book of Agriculture, 1953*

Grape Diseases in the U.S.

During the latter half of the 19th century, grape diseases were among the most damaging and noticeable in the United States as their destructiveness also coincided with an increasingly expanding commercial wine industry. Scribner selected important diseases of the grape and the development of effective control methods as his model. Bordeaux mixture's positive achievements in France corresponded with his efforts to start plant disease research in the U.S., thus this particular fungicide was among the first control methods he attempted.

Black Rot–Scribner's Success Story

At this time in the 1880s American grapevine acres across the country were being abandoned due to one specific disease, known as black rot, caused by the fungus, *Guignardia bidwellii*. Scribner's idea to first study pathogen life cycles proved to be instrumental for determining the fundamental nature of the disease, the understanding of which then directly influenced eventual treatments for it.

Early observations assumed that infections occurring on the fruit and leaves were caused by two distinct fungi. With the assistance of a visiting French viticulture authority, Pierre Viola, Scribner determined that the initial site of infection was on leaves, followed by a fruit rot several weeks later. Treatment with a copper sulfate-lime mixture was then implemented before fruit infection, thereby greatly improving the level of control for the disease. This practice was eventually adopted by producers, reviving the industry, and bringing back production into those sections of the country that were previously deserted.

This became the first great achievement for the USDA, and the second distinct fungal disease to be successfully treated with the fungicide, after late blight. Scribner's investigations with black rot conveniently served as the shining example of his approach to disease control, and the usefulness of the USDA. His success also served as a prototype for conducting future USDA studies in plant pathology.

By the end of 1886, it became obvious that optimal control of plant diseases across the U.S. would require a more systematic approach under the widely disparate regional soil and climatic conditions experienced in a country as large as America. The next year the USDA was able to establish multiple special stations to conduct research on grape diseases based on past problems with black rot, including vineyards in Virginia, New Jersey, Missouri, and Texas. This paradigm is still employed today with USDA-ARS scientists located at various stations around the country working on specific regional or commodity-based problems. ***Now you know the rest of the story.***

Frank Lamson-Scribner spraying the Bordeaux mixture on grape-vines as part of the first plant pathology experiments conducted in the U.S.
Photo Credit: USDA Systematic Botany and Mycological Laboratory, Beltsville, MD.

References

Campbell, C. L., Peterson, P. D., and Griffith, C. S. 1999. The formative years of plant pathology in the United States. APS Press, American Phytopathological Society, St. Paul, MN, 427 pp.

Griffiths, C. S., Peterson, P. D., and Campbell, C. L. 1994. The origins of plant disease research in the Unites States Department of Agriculture. Plant Dis. 78: 318-321.

Peterson, P. D., Griffith, C. S., and Campbell, C. L. 1995. Treatment of grape diseases: the science and successes of Frank Lamson-Scribner. Plant Dis. 79: 89-94

CHAPTER 12

E. F. Smith and the Fusarium Wilts

Introduction

Most of you are likely familiar with E. F. Smith and his seminal work with bacterial diseases in plants, often being referred to as the "Father of American Phytobacteriology". However, prior to his discoveries in bacteriology, did you know that he also pioneered the early studies with Fusarium wilt? That fact is the rest of the story.

Fusarium Wilt Diseases

The genus *Fusarium* was first created in 1809 by Johann Heinrich Friederich Link when he described *F. roseum* as the first species. One of the most important and ubiquitous species, *Fusarium oxysporum*, is the cause of vascular wilt diseases on many plant species. The vascular wilts are considered to be among the most economically important diseases due to their destructiveness and the wide diversity of plants they affect.

George F. Atkinson described the first Fusarium wilt disease in 1892, which had been causing serious losses to cotton growers in Alabama for several years prior to that. He demonstrated the presence of gummy substances blocking the vascular elements and illustrated the distinctive, short conidiophores producing microconidia, now considered to be a primary morphological diagnostic feature for identification of this species.

E. F. Smith's Studies with Fusarium Wilt Diseases

There are other early examples of crops threatened by the Fusarium wilts, including potatoes, celery, flax, bananas, cabbage, and tomato. But the first comprehensive investigations into these diseases were initially conducted by E. F. Smith, beginning with the watermelon. He first described watermelon wilt in 1894 from Georgia and South Carolina. By the time of his second report in 1899, the disease had been detected, spread widely throughout the melon-growing areas of the South ranging from Texas to Maryland, causing enormous economic losses. Large tracts of land, previously cropped profitably to watermelons, were abandoned and rendered unfit for production.

Smith's research on watermelon wilt was combined with a wilt of cowpea and extended work on the wilt disease of cotton previously described by Atkinson. He gave the 3 fungi causing these diseases different names: *F. niveum*, *F. tracheiphila*, and *F. vasinfectum* for watermelon, cowpea, and cotton, respectively based on the fact that they could not be distinguished morphologically, but could be differentiated pathogenically after performing reciprocal cross inoculations on each crop host. Additionally, his broad knowledge of microscopy and plant anatomy allowed him to observe the pathogens' presence and activity within infected plant vascular systems.

He further observed that land infested with the melon fungus should not be planted with watermelons but could be grown with cotton or cowpea. He also demonstrated that the pathogen survived for long periods in the soil and could be spread to new areas on seeds, in run-off water, barnyard manure, and in soil clinging to farm equipment.

Concluding Remarks

Much of what we know today of the fundamental characteristics of the Fusarium wilt diseases was first elucidated by E. F. Smith well over a century ago (125+ years). His investigations into these diseases were critical for those following his lead and helped bring attention to American plant pathology worldwide, giving it more credibility. As the director of the plant pathology laboratory within the USDA's Bureau of Plant Industry, Smith passed the mantle to a brilliant and eager young scientist named William Orton, appointing him the responsibility of solving the Fusarium problem. However, Smith was the first to recognize the importance and potential danger posed by the Fusarium wilt diseases, laying the foundation for others to build upon. ***Now you know the rest of the story.***

E. F. Smith cutting sections of Fusarium wilt-infected watermelon roots. *Photo credit: Special collections, National Agricultural Library, Beltsville, MD. Figure 3.3: in: The Formative Years of Plant Pathology in the United States, 1999 (APS Press).*

References

Campbell, C. L., Peterson, P. D., and Griffith, C. S. 1999. The formative years of plant pathology in the United States. APS Press, American Phytopathological Society, St. Paul, MN, 427 pp.

Harveson, R. M. 1999. Evaluation of the parasitic relationship of *Melanospora* and other allied genera with *Fusarium oxysporum*. Ph.D. thesis. University of Florida, Gainesville, 129 pp.

Harveson, R. M. 2015. The bacterium of many colors. APS Press, St. Paul, MN, 288 pp.

Smith, E. F. 1894. The watermelon disease of the South. American Assoc. Adv. Sci. Proc. 43: 289-290.

CHAPTER 13

Nematodes as Plant Pathogens

Nematodes have been highly successful parasites. Knowledge and awareness of their effects are likely as old as human history, particularly those known to affect animals and humans. Among the earliest records of nematodes are references to the Guinea worm, which invaded the arms and legs of the human body, resulting in severe pain and inflammation.

Records have been found from Egypt as early as 1553-1550 BC describing a large roundworm thought to be Guinea worms, and some historians believe that the "fiery serpents" attacking Israelites mentioned in the Bible (Numbers 21: 6-9) may also have been Guinea worms. In approximately 400 BC, Hippocrates, also referred to roundworm parasites in humans that are thought to be the large intestinal worm, Ascaris.

The smaller nematodes (free-living and plant pathogen-ic) were not noted until microscopes were improved in the 17th century. The first known free-living nematode described was the vinegar eel by Borellus in 1656. Vinegar eels are non-parasitic and feed on bacteria and fungi in unpasteurized vinegar. Although the plant pathogens are often thought to be soil-residing root-infectors, did you know that the first nematodes observed causing disease in plants were those attacking above-ground parts of the host plants? Their story is the rest of the story.

Wheat Gall Nematode

According to Gerald Thorne, the first definitive record of a plant parasitic nematode is assumed to be in William Shakespeare's Love's Labor Lost, in 1594, where he wrote "Sowed cockle, reap'd no corn." This apparently is a reference to the wheat gall nematode (*Anguina tritici*). At that time, it was well known in England that the abnormal, seed-like galls forming in the seed heads as a result of infection were easily seen, and it was known that new plants would not be produced. This condition was referred to in Britain as purples, pepper corns, or more commonly, cockles. It was also known in Germany as *Rodenkrankheit* (the cockle disease), stemming from the association of the cockles with seeds of the common weed, the cockle (*Agrostemma githago*) that often became mixed with wheat seed during and after harvest.

The English clergyman, Turbevill Needham, has generally been credited as the first to document a plant parasitic nematode in the scientific literature in 1743. However, he never recognized its role in plant pathology. Instead he interpreted his discovery as conclusive proof for the concept of spontaneous generation, for which he strongly believed. He crushed some shrunken, blackened wheat grains and viewed them under a microscope in a drop of water, and was amazed to view a mass of dried inanimate fibers that suddenly came to life and began to twist and move.

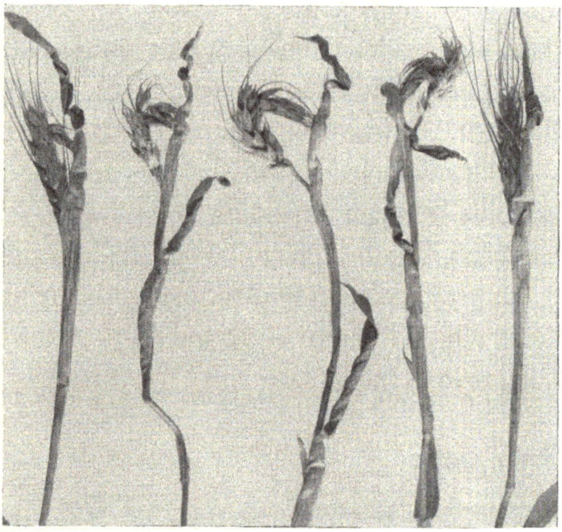

Symptoms on wheat plants infected by the seed gall nematode (*Anguina tritici*). *Photo credit: R. W. Leukel, USDA.*

Davaine's Contribution

Several other workers followed with additional studies, but none contributed substantially to the knowledge base of the wheat gall disease until the physician Casimir Davaine in the mid-19th century. More details of Davaine's work with both animal (anthrax) and plant bacteriology have been covered in another chapter of this book. However, his most lasting contribution to science may have been his studies with the wheat gall nematode. His pioneering experiments greatly improved the understanding of the nematode's biology and life cycle. He recognized and described the complete developmental growth stages of the pathogen during the infection process, which played a major role in suppressing scientists that were maintaining and guarding the concept of spontaneous generation.

Stem and Bulb Nematode

The German plant pathologist, Julius Kuhn discovered another above-ground endoparasitic nematode infesting the heads of Fuller's teasel (*Dipsacus fullonum*), and published his results in the same year (1857) that Davaine reported his findings. That nematode is now known as the stem and bulb nematode, *Ditylenchus dipsaci*.

Symptoms of the disease were actually observed two years earlier by Schwertz in a rye field. He named this condition "stock", in reference to the stiff, baton-like plant stems, characteristic of the disease, but never observed the nematode itself on affected plants. Kuhn found a nematode infecting the heads of teasel, naming it *Anguillula dipsaci* (species epithet based on the host genus *Dipsacus*) and further noted that the nematodes causing damage in rye were similar to the teasel pathogen. He then reproduced the disease in rye by placing nematode-infested teasel heads within the rows of the grain, thereby providing proof of pathogenicity in both crops by the same nematode.

Conclusions

It is not particularly surprising that the first nematode plant pathogens recognized were found infecting plant foliage and stems in the mid-1850s. Most of the early diseases described by that time (mildews, rusts, etc.) also were easily detected while causing foliar blights and leaf spots. Few researchers had yet investigated underground sources of plant problems. Soilborne root diseases that seriously threatened the sugar beet industry (sugar beet cyst and Rhizoctonia root rot), would soon be recognized and closely studied by Kuhn and others in central and northwestern Europe. ***Now you know the rest of the story.***

References

Barker, K. R. 2004. A century of plant nematology. Pages 1-51 in: nematology – advances and perspectives. Volume I: nematode morphology, physiology, and ecology. Z. X. Chen, S. Y. Chen, and D. W. Dickson, eds. Tsinghua University Press, 636 pp.

Christie, J. R. 1959. Plant nematodes: their bionomics and control. H. and W. B. Drew Company, Jacksonville, FL. 256 pp.

Harveson, R. M. 2015. The bacterium of many colors. APS Press, St Paul, MN, 288 pp.

Thorne, G. 1961. Principles of nematology. McGraw Hill Book Company, New York, 553 pp

CHAPTER 14

WWI Influences the Birth of Fumigation and Nematology

At the start of the First World War, the major powers were still immersed in the warfare traditions of the 19th century, and expected to fight it in much the same way as previous conflicts. Troops would engage in a few pitched battles, after which the loser would rapidly plea for peace. Instead, two evenly matched sides found themselves in a battlefield stalemate, locked in an exhausting war of attrition. Due to desperation, remarkably ingenuous new ways to kill were invented and used for the first time in the effort to gain advantage, including machine guns, long-range artillery, aircraft, tanks, and poison gases.

In spite of the horrendous and futile loss of life over the course of the four-year war, several unanticipated after-effects emerged that elevated plant pathology. The knowledge obtained from studying and employing poison gases during the war catalyzed the concept of soil fumigation for disease control, which in turn revealed the previously unknown importance of nematodes in agricultural production. That is the rest of the story.

Nematodes

Nematodes have presumably played an important but often unrecognized role in agriculture, where they are thought now to have been at least partly responsible for the rapid decline of cultivated soils in the colonies along the Atlantic coast. It is also thought that this reduction in productivity was one of the incentives for the migration of early pioneers westward searching for better farming conditions in the early to mid-1800s.

Nematologist Gerald Thorne later analyzed many of these soils in the 1930's, and identified multiple plant pathogenic species belonging to root-knot, spiral, and lesion nematode groups. He remarked that it is likely that the pathogenic species were native to the soils, but had simply become problematic on introduced crops after their indigenous hosts were destroyed for agricultural purposes.

Fumigation

The concept of soil fumigation was initiated in the 1860s utilizing carbon disulfide for managing the root-infecting insect (Phylloxera) that was devastating French vineyards. Due to the success of this method for controlling insects, it was unsuccessfully attempted for controlling sugar beet cyst nematodes in Germany in 1871. When carbon disulfide proved too costly for this purpose, a number of bizarre and fruitless soil amendments were eventually explored, including human urine and extracts from decomposing cadavers.

In the U.S. in the early 1900s, other chemicals such ammonium sulfate, formaldehyde, cyanide, and quicklime were all tested as potential nematicides. However they were all too expensive. It was not until after WWI that any soil fumigant was demonstrated to be economically effective.

Chloropicrin

During WWI, chloropicrin (trichloronitromethane) was among the major chemical warfare agents used by the U.S. military, with chlorine, mustard gas, phosgene, and Lewisite. It was not a new and unknown compound even at that time. It had been first synthesized in 1848 by the Scottish chemist, John Stenhouse. Chloropicrin was a lacrimator (tear producer) and severe respiratory irritant that often caused vomiting but was not generally fatal. Its purpose was to force nauseated soldiers to remove gas masks, exposing them to more lethal gases that were often loaded into artillery shells with chloropicrin.

After the war was over, the nematicidal properties of chloropicrin were discovered in England, and large quantities of the product were declared surplus and readily available. It was then tested as a fumigant in locations where poor yields followed continuous culture, particularly vegetable and fruit production areas of California and Hawaii. In California it was shown to kill 99-100% of nematode populations after being injected 6-8 inches deep and tarped to reduce loss due to vaporization.

In the late 1920's, work by M. O. Johnson and G. H. Godfrey produced spectacular results in yield improvements and disease reduction against root-knot nematodes in Hawaiian pineapple fields. This practice became the standard in Hawaii and its use continued in the pineapple industry until supplies of chloropicrin were exhausted.

Various types of gas masks utilized in World War I.
Photo credit: Library of Congress

Conclusions

The process of fumigation made it possible to demonstrate that certain disease symptoms formerly attributed to fungi and bacteria were actually due to plant parasitic nematodes after they were eliminated from the soil. Prior to this it had been difficult to quantify crop losses attributed specifically to nematodes because of their soil residence and common interactions with other soilborne pathogens. It also spurred the creation of a new discipline (nematology) within plant pathology and a novel method of disease control–both positive outcomes derived from the chemical warfare of the First World War. ***Now you know the rest of the story.***

References

Christie, J. R. 1959. Plant nematodes: their bionomics and control. H. and W. B. Drew Company, Jacksonville, FL. 256 pp.

Harveson, R. M. 2015. The bacterium of many colors. APS Press, St Paul, MN, 288 pp.

Huber, D. M., and Schneider, R. W. 1982. The description and occurrence of suppressive soils. Pages 1-7 in: suppressive soils and plant disease. Schneider, R. W., ed. American Phytopathological Society, St. Paul, MN.

Medical Aspects of Gas Warfare. 1926. From: The medical department of the United States in the World War, Volume XIV, prepared under the direction of Maj. Gen. M. W. Ireland, Washington: Government Printing Office.

Thorne, G. 1961. Principles of nematology. McGraw Hill Book Company, New York, 553 pp.

Tucker, J. 2006. War of Nerves: Chemical Warfare from World War I to al-Qaeda, Pantheon Books, New York. 482 pp.

CHAPTER 15

Root-Knot Nematode Improves Fumigation

The inception of fumigation with chloropicrin after WWI and its role in illustrating the previously unknown importance of nematodes to crop production has been introduced in a previous chapter of this book. The root-knot nematode specifically is thought to be directly responsible for the evolution and development of chemical fumigants in the 20th century, while also initiating the study of nematology. That is the rest of this story.

Root-Knot Nematode

The name "root-knot" refers to distinctive root galls associated with infection by this nematode. These species are endoparasites that complete most of their life cycle within the roots of their hosts. They survive in soil primarily as eggs, and infect host plants after hatching as second stage juveniles. The parasite becomes sedentary after infecting roots, and permanent feeding sites are established, providing nutrients as the nematode continues to feed and enlarge. Root cells around the feeding sites also enlarge and form the characteristic galls (knots).

What we now know as the root-knot nematode was first described by Miles Berkeley England in 1855 from roots of cucumber plants grown in a greenhouse. The first reports of root-knot nematode in the U.S. were made from Alabama and Florida by G. F. Atkinson and J. C. Neal, respectively in 1889. Atkinson further determined that root-knot infections in cotton promoted and enhanced Fusarium wilt infections thereby introducing the concept of multiple root diseases occurring simultaneously. Atkinson's results also brought attention to this problem in the U.S. and generated new research interests within the scientific community.

Gall symptoms of root-knot infections on sugar beet roots. *Photo credit: R. M. Harveson*

Most of the early knowledge pertaining to this nematode came from European workers in the last half of the 19th century. These studies tended to concentrate on various aspects of the nematode itself (morphology, life cycle, and ecology), whereas the majority of the American studies focused on control, rather than the nematode and its biology.

Fumigation Emerges in the U.S.

Although fumigation was first attempted in response to the grape Phylloxera in French vineyards, the emergence of the root-knot nematode strongly influenced the search for and development of new and improved soil chemicals for disease control in the U.S. Previous management experiments with trap crops and crop rotation were of limited value. Thus an emphasis was placed on the identification and development of new soil fumigants.

Methyl Bromide (Bromomethane)

Methyl bromide, like chloropicrin, was introduced into the United States in the 1920s. It was used successfully as a soil fumigant for nematodes and insects into the late 1930s after the WWI surplus chloropicrin supplies were gone. One of its disadvantages was a need to be evaporated under a tarp cover due to its properties of transforming into a gas with warming soil temperatures. It also tended to cause phytotoxicity unless planting was delayed two to four days after application.

New Chemical Fumigants

The outstanding success of chloropicrin and methyl bromide for nematode control stimulated the search for new and better chemical products. In 1943, Walter Carter, an entomologist with the Hawaiian Pineapple Research Institute reported that D-D (an equal mixture of 1,3-dichloropropene and 1,2-dichloropropane) effectively controlled both root-knot nematodes and the destructive root-feeding larvae of the Oriental beetle (*Anamola orientalis*).

Soil fumigants on a field scale in the 1950s. Note the lack of masks or protective clothing. *Photo credit: Florida State Plant Board (DPI).*

The discovery of another chemical known as ethylene dibromide (EDB) as an effective agricultural fumigant soon followed, and the modern development of nematicides was underway. These volatile chemicals produced spectacular improvements in growth and increased yields in response to root-knot infection, and several advantages were noted from their use. Both D-D and EDB were by-products of the petroleum refining industry therefore their price was now less expensive due to the lower cost of producing the chemical compounds. These products were additionally found to be safer and easier to handle than chloropicrin while also producing similar results.

Conclusions

The successful introduction of D-D by the Shell Chemical Corporation in 1944, and EDB by the Dow Chemical company shortly after, further generated widespread interest in these types of products for nematode control worldwide. By the 1950's, economical control became possible with the advent and discovery of many new chemicals for use as soil fumigants.

Efforts to control the root-knot nematodes brought to light the importance and damaging effect of nematodes in general and root-knot in particular to agriculture throughout the world. Furthermore, by the 1960s, serious study of nematodes was underway. Land grant universities began to emphasize nematology as a distinct discipline, which created the opportunity for new research funding aimed at managing an increasingly important problem. ***Now you know the rest of the story.***

References

Barker, K. R. 2004. A century of plant nematology. Pages 1-51 in: Nematology – advances and perspectives. Volume I: nematode morphology, physiology, and ecology. Z. X. Chen, S. Y. Chen, and D. W. Dickson, eds. Tsinghua University Press, 636 pp.

Christie, J. R. 1959. Plant nematodes: their bionomics and control. H. and W. B. Drew Company, Jacksonville, FL. 256 pp.

Harveson, R. M. 2015. The bacterium of many colors. APS Press, St Paul, MN, 288 pp.

Taylor, A. L. 2003. Nematocides and nematicides – a history. Nematropica 33: 225-232.

Thorne, G. 1961. Principles of nematology. McGraw Hill Book Company, New York, 553 pp.

CHAPTER 16

The Native Nebraska Nematode

Introduction

The devastating consequences resulting from accidentally importing exotic pathogens into new areas has long been known and well documented in plant pathology history. Examples include the introduction of the pathogens causing the chestnut blight and Dutch elm diseases and their disastrous effects upon forest and shade trees in North America and Europe.

Conversely, there are a number of endemic plant pathogens in a given area that were discovered only after they caused economic damage on newly introduced host plant species. This article will focus on that topic using the old-world crop, the sugar beet, as the example.

The current sugar beet is a product of breeding research. It was derived from the domesticated beet (*Beta vulgaris*), which was, in turn, previously developed via selections made from the wild beets native to Mediterranean Sea coasts and adjacent Northern Europe coastal areas. In Europe, the crop had been notoriously long known as highly susceptible to diseases from areas into which it was introduced, such as Rhizoctonia root rot and the cyst nematode. In the Americas, we are now aware of a nematode pathogen never before documented prior to the arrival of the sugar beet into the new world. That nematode, known as the false root-knot nematode, is the rest of the story.

False Root-Knot Nematode

The false root-knot nematode (*Nacobbus aberrans*) has also been referred to in the literature as the Nebraska root galling nematode. These nematodes produce galls on the roots of their hosts that look similar to those caused by several species of the root-knot nematodes (*Meloidogyne* spp). If diagnoses were based on symptoms alone, they could easily be mistakenly attributed to *Meloidogyne*, thus these nematodes were then referred to as false root-knot nematodes.

This nematode was discovered and first described by Gerald Thorne in 1935 on non-agricultural land from the native desert shrub, shadscale (*Atriplex confertifolia*), near Utah Lake, Utah. It was placed into a newly created genus in 1944 (*Nacobbus*) with two species, *N. dorsalis* and *N. aberrans*. This new genus was made in honor of Nathan Augustus Cobb, the pioneering USDA pathologist who is commonly cited as the "Father of American Nematology".

New Sugar Beet Disease in North America

The false root-knot nematode was first reported affecting sugar beets in 1949 near Mitchell, NE, by University of Nebraska plant pathologist, Max Schuster. However, the disease symptoms had been previously recognized for years, but mistaken for root-knot nematode infections. The most conspicuous symptoms of the disease are galls or swellings with a proliferation of lateral branches formed on mature beet taproots. These lateral roots may be studded with galls, giving them a beaded appearance.

After thorough investigations of distribution, host range, life cycle, and pathology were completed, the sugar beet pathogen from Nebraska was interpreted to be distinct from either of the two previously mentioned species (*N. dorsalis* and *N. aberrans)* and named *N. batatiformis*, based on the shape of the mature female. At maturity, the female's body was tapered at both ends, resembling a sweet potato (*Ipomoea batatas*). However, this species was later synonymized with *N. aberrans*, retaining that name.

Sugar beet plants infected by the False Root-Knot nematode at Harvest. Note the branched, lateral branches studded with galls resembling a beaded necklace or chain. *Photo credit: R. M. Harveson*

Distribution and Host Range

A regional survey from the late 1950s determined that the false root-knot pathogen was widespread throughout the Great Plains states, infesting one third (32%) of 125 sugar beet fields surveyed in Nebraska. Today, infestation of sugar beet fields is limited to the High Plains east of the Rocky Mountains, yet still considered a potentially severe pest due to previous documentations of sugar beet losses approaching 20%.

In the United States, *N. aberrans* is an economic problem primarily on sugar beet, but it can also affect numerous other vegetable crops. Fortunately, most field crops grown in Nebraska and the Central High Plains under irrigation, other than sugar beet, are not susceptible, including potatoes, corn, dry beans, sunflower, alfalfa, wheat, and barley. Because of the broad host range, the disease can also be problematic for city residents with backyard gardens, particularly with root vegetables like beets, turnips, radishes, and carrots.

Interestingly, Schuster also discovered this nematode on three species of cacti, including two species of the prickly pear, and the spiny star cactus, all of which are native to western Nebraska. This pathogen has also been found commonly infecting native weed species within fields containing various mustards, common lambsquarters, and Kochia as well as introduced species such as common purslane, puncturevine, and Russian thistle.

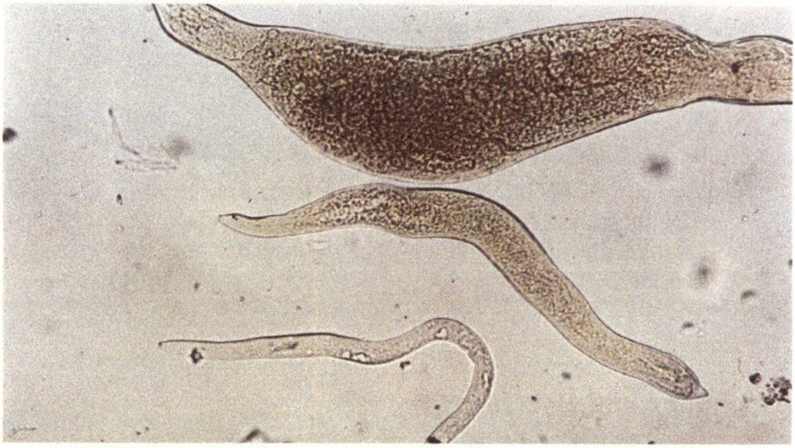

The sweet potato-shaped body of a mature female false root-knot nematode with various stage forms of juveniles. *Photo credit: M. L. Schuster, reproduced from R. M. Harveson et al. 2013. Sugar Beet Production Guide. University of Nebraska, Lincoln, Cooperative Extension Circular 156.*

Uncertain Taxonomy

The taxonomic history of *N. aberrans* has been controversial and complex. However, it is still generally accepted that this nematode can be separated into at least three pathotypes or groups based on different DNA characteristics and host preferences. These pathotypes include: 1) a strain from the highland Andean regions of Argentina, Bolivia, Chile, and Peru that attacks potatoes but not peppers; 2) a strain from Mexico that damages beans, tomatoes, and peppers, but not sugar beet or potato; and 3) the North American strain that infects sugar beets and many common vegetable crops, with the exception of potatoes, throughout the Great Plains of the U.S.

Conclusions

Irrespective of the confusion existing with the current taxonomy of this nematode, it can now be clearly concluded that the genus *Nacobbus* (false root-knot nematode) is distinct from the various *Meloidogyne* species (root-knot nematodes) with which it has often been confused, and is also unquestionably native to the Americas. Strains of it have been identified from highly diverse climatic areas of the Western Hemisphere, yet absent from any other location in the world. Furthermore, the documentation of the sugar beet strain (formerly *N. batatiformis*) infecting native cacti from uncultivated rangeland by Schuster in the 1950s, provided strong evidence that this nematode was indigenous to western Nebraska, but adopted the sugar beet as its preferred host after it was introduced into North America. ***Now you know the rest of the story.***

References

Harveson, R. M. 2008. False root-knot nematode. NebGuide G1857, Coop. Ext. Serv., Univ. of Nebr., Lincoln.

Harveson, R. M., Panella, L., and Lewellen, R. T. 2009. History of beet production and usage. Pages 1-2 in: Compendium of Beet Diseases and Pests. R. M. Harveson, L. E. Hanson, and G. L. Hein, eds. APS Press, St. Paul, MN.

Harveson, R. M. 2015. The bacterium of many colors. APS Press, St Paul, MN, 288 pp.

Thorne, G. 1935. The sugar beet nematode and other indigenous nemic parasites on shadscale. J. Agric. Res. 51: 509-514.

Thorne, G. and Schuster, M. L. 1956. *Naccobus batatiformis*, n. sp. (Nematode: tylenchidae), producing galls on the roots of sugar beet and other plants. Proc. Helminthol. Soc. Wash. 23: 128-134.

Schuster, M. L., and Thorne, G. 1956. Distribution, relation to weeds, and histology of sugar beet root galls caused by *Naccobus batatiformis*, Thorne and Schuster. J. Am. Soc. Sugarbeet Technol. 9: 193-197.

Schuster, M. L., Sandstedt, R., and Estes, L. W. 1965. Host-parastie relations of *Naccobus batatiformis* and the sugar beet and other hosts. J. Am. Soc. Sugarbeet Technol. 13: 523-537.

CHAPTER 17

The First Plant Viruses?

In the late 1880s, fungi and bacteria were gradually being recognized as causal agents for plant diseases, but this was not universally accepted by the scientific community. At the same time a number of infectious diseases were being observed and described where no corresponding biological cause could be identified. Tobacco mosaic was one of the first of these unknown infectious plant diseases to be recognized as economically damaging and extensively studied. However did you know that other diseases now known to be caused by viruses were noted much earlier? Pictorial and written records exist that provide clear evidence that plant virus infections were recognized, although it is equally obvious that these individuals had no concept or understanding of what they were observing. One well-known example of this is the tulip-breaking disease, caused a potyvirus (*tulip-breaking virus*).

Tulip-breaking

"Breaking" is a term for the unusual but beautiful variegated color patterns, consisting of stripes, flecks, and featherings developing in tulip flower petals. Those plants exhibiting these symptoms became highly desired and extremely valuable. Demand for broken tulips soon exceeded supplies, creating a wild speculation craze, referred to as Tulip Mania, reaching its peak between 1634 and 1637. Fortunes were won and lost when the market eventually collapsed after it was discovered that the breaking color patterns could be produced by grafting the mosaic bulb type to a common type. Also contributing to the decline of tulip-mania was the observation that the vigor and bulb quality substantially diminished from broken tulips several years following infection. It was not until 1929 that the cause of the breaking was shown to be caused by viral infection. Two additional examples that are likely much more unfamiliar to readers include eupatorium yellow vein and nettlehead. The discovery, history, and investigations of these somewhat obscure diseases are the rest of the story.

Symptoms on tulips of *Tulip mosaic virus*.
Photo Credit: Wikipedia Commons.

Eupatorium yellow vein

To our knowledge today, the earliest known reference to a plant virus was recorded by the Japanese Empress Koken in 752 AD from the oldest poem collection in Japan (Man'yoshu), consisting of more than 4500 writings. The poem mentions the unusual appearance of bright yellow leaf veins forming on leaves of *Eupatorium chinense*, an ornamental flowering plant in the aster family. She apparently noticed the characteristic yellow vein clearing patterns during the summer. This observation was novel because such patterns normally occurred in fall as the plant senesced. It has recently been proven that a geminivirus (*Eupatorium yellow vein virus*) in combination with a DNA β-satellite component are responsible for the characteristic foliar patterns on eupatorium plants.

Nettlehead and Discovery of Nematode-Transmitted Nepoviruses

Nettlehead is a hops disease, first described in a book written in 1574, entitled "A Perfite Platform of a Hoppe Garden" by a British landowner named Reginald Scott. It still remains one of the most destructive diseases of this crop inducing progressively severe symptoms followed by death 5-6 years after initial infection.

Early reports indicated that the most serious outbreaks occurred in hop rows planted along former hedgerows and after cultivation of new pastures. This suggested a soilborne relationship, but many years elapsed before the dagger nematode (*Xiphinema diversicaudatum*) was identified as the vector.

In the 1960s K. R. Bock found polyhedral (spherical) viral particles present in all nettlehead-diseased plants and identified them as *Arabis mosaic virus* (*ArMV*). Two decades later, several low molecular weight satellite nucleic acid species were consistently found in association with *ArMV* in infected hop plants, proving that a second factor was involved since all diseased plants with nettlehead symptoms yielded *ArMV*, but not all *ArMV*-infected plants developed the nettlehead disease.

We now know definitively that nettlehead is caused by the hop strain of (*ArMV*), a nepovirus, in association with at least two satellite nucleic acids. It is serologically similar to *grapevine fan leaf virus* (*GVFLV*), also a nepovirus and the first recorded plant virus demonstrated to be transmitted by nematodes (several species of *Xiphinema*). ***Now you know the rest of the story.***

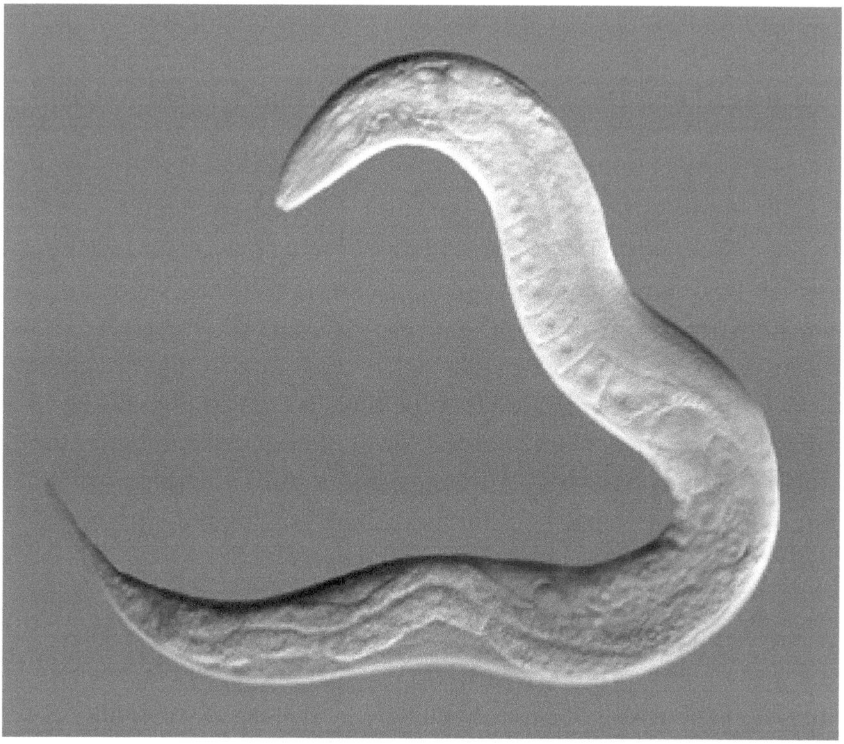

Dagger nematode.

Credit: Wikipedia, Bob Goldstein, UNC Chapel Hill http://bio.unc.edu/people/faculty/goldstein/ - Own work

References

Davies, D. L., Clark, M. F. 1983: A satellite-like nucleic acid of *aribis mosaic virus* associated with hop nettlehead disease. Annals of Appl. Biol. 103: 439-488.

Inouye, T., and Osaki, T. 1980. The first record in the literature of the possible plant virus disease that appeared in "Manyoshu", a Japanese classic anthology, as far back as the time of the 8th century. Ann. Phytopathol. Soc. Jap. 46: 49-50.

Saunders, K., Bedford, I. D., Yahara, T., and Stanley, J. 2003. Aetiology: the earliest recorded plant virus disease. Nature 422: 831.

Stevens, W. A. 1983. Virology of flowering plants. Blackie, Glasgow and London. 183 pp.

CHAPTER 18

Polymyxa betae and Rhizomania in the U.S.

Since its first report from Italy in the 1950s, rhizomania has been arguably the most feared disease of sugar beets, causing severe damage wherever this crop is grown. This disease is also very unusual due to the fact that it is a soilborne, root rot caused by a virus known as *beet necrotic yellow vein virus* (*BNYVV*). The pathogen is uniquely transmitted to host roots by a zoosporic, plasmodiophorid parasite, *Polymyxa betae*. The natural host range of the parasite is limited to sugar beets, spinach, chard and related weed species such as pigweed, lambsquarters, and Kochia. Without the soil-inhabiting vector, disease cannot occur, however the parasite does not always contain the virus. The vector itself causes little damage unless it is harboring the virus, transmitting it to the plant roots after infection.

New Reports

Due to the fear of this disease and the potential damage to sugar beet production, my predecessor, Eric Kerr (University of Nebraska, Panhandle REC, Scottsbluff) and Willem Langenberg (USDA-ARS, UNL-Lincoln), proactively conducted a survey in 1976 from western Nebraska sugar beet fields searching for root infections by the vector, but found none. To the best of my knowledge, this was the first attempt to identify the vector/pathogen/disease system in sugar beet production in the Americas.

The following year, *Polymyxa betae* was detected for the first time in the Western Hemisphere by Bryce Faulk and James Duffus, (USDA Salinas, CA), although it was not found directly infecting sugar beet roots in the field. It was identified in the greenhouse from beet plants grown in soil samples collected from two California fields where previous beet crops had exhibited yellowing symptoms. Furthermore, no evidence of the virus or symptoms of the disease (rhizomania) was observed, only the presence of the vector (*P. betae*) in the baited plants from the field soil samples.

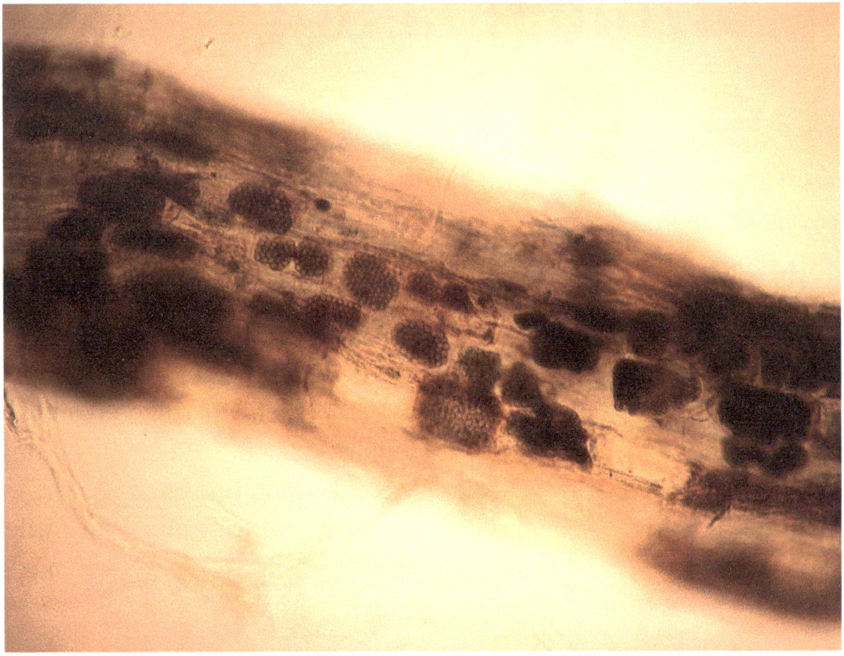

Small, infected sugar beet rootlet infected by *Polymyxa betae*. The small circular cysts will release zoospores that carry and reinfect roots with the viral pathogen. *Photo Credit: R. M. Harveson*

Curiously, the virus pathogen (BNYVV) was first encountered in North America in 1981 from a cherry orchard in Washington State, from a location having nothing to do with sugar beet production. In the attempt to correlate a potential soilborne virus with the stem-pitting disease of sweet cherry, Abdullah Al Musa and Gaylord Mink identified *BNYVV* from the roots of *Gomphrena globosa* (globe amaranth) grown in soil samples collected from beneath diseased cherry trees.

Due to these new reports of both the vector and virus from California and Washington respectively, Kerr and Langenberg conducted another search in 1981, examining sugar beet and redroot pigweed plants from low-lying areas of sugar beet fields in Scotts

Bluff County, Nebraska. In five of six fields inspected, *P. betae* was present in the roots of all plants that were collected and tested. This was the first evidence showing the vector directly infecting sugar beets in the field from North America. However, the virus causing rhizomania was not detected from this search. The pathogen and disease would not be found affecting sugar beets in Nebraska for another decade. Ironically, I was involved with that first detection of rhizomania in Nebraska and several other states, and that is the rest of the story.

Sugar beet roots at harvest infected with the viral pathogen beet necrotic yellow vein virus. *Photo Credit: R. M. Harveson*

My Unlikely Role

Finding rhizomania in the U.S. was an alarming development to the industry due to its well documented damage elsewhere, and with no management options available. It also placed an unwarranted pariah-like stigma on those unfortunate farms of first identification from each new state or region. Therefore, any new reports were of great importance and concern.

In July 1992, I was working as a research technician in the

program of Charlie Rush with the Texas Agricultural Experiment Station in Bushland, TX, one of only two labs in the western hemisphere that were capable of testing for rhizomania at that time. The other lab with this ability was that of James Duffus, who had first found the disease in U.S. sugar beet production in California in 1983 and Texas in 1985.

I personally ran the tests on sugar beet samples sent to us by Eric Kerr that confirmed the presence of the disease in Nebraska. I also processed additional samples that same summer identifying the disease from Wyoming and Idaho as well. I later learned that our tests were simply a second verification after first being confirmed by Duffus earlier that summer, thus making my role a very minor one. Nevertheless, it was a foreshadowing, landmark event in my career. Never in my wildest dreams could I have imagined in 1992, that I would replace Eric Kerr seven years later as the extension plant pathologist in Scottsbluff-it is indeed a small world. ***Now you know the rest of the story.***

References

Al Musa, A. M., and Mink, G. I. 1981. Beet necrotic yellow vein virus in North America. Phytopathology 71: 773-776.

Faulk, B. W., and Duffus, J. E. 1977. The first report of *Polymyxa betae* in the western hemisphere. Plant Dis. Rep. 61: 492-494.

Langenberg, W. G., and Kerr, E. D. 1982. *Polymyxa betae* in Nebraska. Plant Dis. 66: 862.

CHAPTER 19

The Role of Curly Top in U.S. Sugar Beet Production

Introduction

Elsewhere in this book, I have shared the story of how WWI impacted the development of the sugar beet industry in the US. I also alluded to the fact that in my mind there was another major factor that assisted in the revelation for the need to create a domestic seed production for maintaining consistent and dependable quantities of sugar for American consumers.

After the First World War was over, roughly 15,000,000 pounds of seed were still being purchased and imported from Europe, and this continued into the early 1930s. However, it was a viral plant disease called beet curly top that finally provided the most convincing reason for establishing a sugar beet seed industry in the U.S., completely independent from European sources. Beet curly top did not exist in Europe, thus the varieties from Europe grown in the U.S. were all highly susceptible to the disease. In fact, outbreaks of curly top were so severe throughout the first three decades of the 20th century, that this nascent, struggling industry (sugar beet production) was nearly destroyed before it really began. This is the rest of the story.

Curly Top Emerges

Curly top of sugar beet was among the first plant viral diseases recognized in the late 19th century (with tobacco mosaic), however its early known history in the United States is limited because it shares similar symptoms with other diseases and disorders. According to C. O. Townsend, USDA's George G. Hedgcock reported the observation of damaging outbreaks of a disease from garden beets in Nebraska before 1888 that are now presumed to be from curly top. In the 1890s, severe losses from an unidentified disorder resembling curly top were also reported from Utah and California. By 1900 this problem had been noted throughout all western states where sugar beets were grown.

In 1915 the relationship between the pathogen and its leafhopper vector (*Circulifer tenellus*) was established, suggesting a viral etiology, however the pathogen was not identified for another 60 years. It was not until 1974 that a virus (*Beet curly top virus*) was properly proven to be the cause of curly top, after advancements in virus purification and the twinned particles characteristic of geminiviruses were observed with electron microscopy.

Curled leaf symptoms of a sugar beet plant infected with the curly top pathogen *Beet curly top virus. Photo credit: R. M. Harveson*

Resistance Breeding Begins

Because of successive crop failures from the curly top problem throughout the intermountain west of the United States, the USDA began breeding for disease resistance, and the plant pathologist, Eubanks Carsner, was assigned to this task in 1919. He initiated these new breeding efforts targeting the improvement of sugar beets by selecting individual plants exhibiting resistance within fields with high levels of disease pressure, a system referred to as mass selection. By 1929, Congress had appropriated additional funds for continuing this line of research. These same methods had previously been developed and implemented successfully by William Orton and H. L. Bolley for Fusarium wilt diseases in the first decade of the 20th century.

Success of "US 1"

The testing of seed accumulated between 1925 and 1929 (about 100 pounds) from surviving plants obtained through this method was first conducted in Idaho in 1930. Those surviving plants from these evaluations eventually produced line 'US 1". Results were not particularly outstanding, but compared to European cultivars, the performance of US 1 was considered exceptional. In 1931, based on results from 32 different sites with moderate to severe disease pressure, US 1 averaged more than 5 tons/acre higher (root yields) than the European entries.

Although this new line was immediately recognized as purely a stopgap measure, it was still released as the first curly-top resistant cultivar for commercial production in 1933. Steps were taken immediately to increase seed as a source of resistance to curly top. US 1 was quickly replaced by other superior lines (US 33 and 34), created by reselections from US 1, and both were more resistant than their parents. For example, 85% of planted sugar beet acreage in 1934 was abandoned and the average national yield was 4.9 tons/acre. By 1941, 97% of the acreage was harvested and processed with national yields improving to 13.5 tons/acre. By the mid-1940's, the disease no longer posed a serious problem to the beet sugar industry. For this achievement, Carsner and his staff received the Superior Service Award of the USDA in 1947.

Conclusions

This one disease was highly responsible for the closure of numerous American processing plants in a manner similar to the impact of the cyst nematode on the European continent in the latter half of the 19th century. The development of new resistant cultivars in the 1930s quickly reversed these trends, reduced damage to crops, and is generally credited with saving the industry. It also confirmed the necessity for creating a domestic sugar beet industry, hastening its initiation.

Although crop failures today are rare, curly top is currently still a problematic issue for sugar beet production in several areas of the western U.S. If the curly top problem had not been solved, is it possible that we might be receiving the majority of our sugar from cane? ***Now you know the rest of the story.***

References

Carsner, E. 1924. Studies on curly top disease of the sugar beet. J. Agric. Res. 28: 297-320.

Carsner, E. 1927. Sugar beet disease called curly top limits production. Pages 603-605 in: yearbook of agriculture.

Coons, G. H., Owen, F. V., and Stewart, D. 1955. Improvement of the sugar beet in the United States. Advances in Agronomy 8: 89-135.

Harveson, R. M. 2015. The bacterium of many colors. APS Press, St Paul, MN, 288 pp.

Panella, L., Lewellen, R. T., and Harveson, R. M. 2009. Breeding for disease and insect resistance. Pages 3-5 in: Compendium of beet diseases and pests. R. M. Harveson, L. E. Hanson, and G. L. Hein, eds. APS Press, St. Paul, MN.

Townsend, C. O. 1917. The present status of the sugar-beet seed industry in the United States. U. S. Dept. Agr. Yearbook 1916: 399-410.

CHAPTER 20

Rice Dwarf Virus

The First Plant Virus Known to be Insect Vectored

Introduction

At the turn of the 20th century, viruses were just becoming recognized as causal agents for plant diseases. In a previous chapter, I shared the story of some of the first plant virus diseases noted long before the early 1900s, but not understood–eupatorium yellow vein, tulip-breaking, and nettlehead in hops. Although not recognized as early as other virus diseases discussed previously, *Tobacco Mosaic Virus* (*TMV*) is universally considered the first (plant or animal) virus to be thoroughly studied. As the disease was gradually investigated and characterized in the late 1890s, it became the standard by which all subsequently discovered plant viruses were measured.

TMV was demonstrated to be the first pathogenic agent smaller than bacteria. It furthermore was instrumental in the establishment of general concepts of virology and the development of molecular biology after its structure was identified to be composed of both protein and nucleic acid components. Most of this is probably familiar with veteran plant pathologists, but did you know that another lesser-known virus disease was being extensively studied in Japan simultaneously and/or even earlier? That fact is the rest of the story.

Rice Dwarf

The highly destructive virus disease, rice dwarf was apparently recognized as early as 1883 in Japan, making it a lesser-known contemporary with tobacco mosaic. By 1892, it had spread widely throughout the country, affecting more than 1100 acres of rice fields. The relationship with leafhoppers and disease was established by 1894, and it was determined that the disease was not caused by the insect, but some other unknown agent carried by the leafhopper. After a series of experiments were conducted between 1897 and 1906 at the Shiga Prefecture Agricultural Experiment Station, it was formally demonstrated that rice dwarf was caused by a virus and disseminated to rice plants by the leafhopper, *Nephotettix apicalis* var. *cincticeps*. These experiments further determined that virus transmission was not achieved by seeds, soils, or mechanically through infected plant sap, but that the disease was effectively managed by eliminating the insect vector. They also presciently noted a long incubation period was required for transmission by the insects, and uninfected leafhoppers became infected only after feeding on diseased plants for 15 days.

Today we recognize *Rice dwarf virus* (*RDV*) as one of three members of the genus *Phytoreovirus*. The double stranded, RNA pathogen is transmitted exclusively by leafhoppers in a persistent manner, requiring replication within its vector, thus a latent period of about 2 weeks is required between acquisition and transmission. This now explains and confirms the astute observations from the Japanese researchers at the turn of the 20th century.

This ground-breaking work was significant for a number of reasons. Rice dwarf proved to be the first virus disease demonstrated to be transmitted by an insect, as well as the first one found affecting cereals. It also held the distinction of being the first shown to be transovarial (pathogen passing from the parent through the egg to the next generation's offspring), and to multiply and be viewed within its vector by electron microscopy.

Concluding Remarks

Tobacco mosaic proved to be unique among plant viruses as no biological vector was ever identified or even needed for its transmission. As new virus diseases continued to be brought to light, it became apparent that insects were an important factor and the primary mechanism for the transmission and spread of many of these diseases. In fact, most plant viruses since that time have been demonstrated to be transmitted by insects, or some other organism, including rice dwarf virus. ***Now you know the rest of the story.***

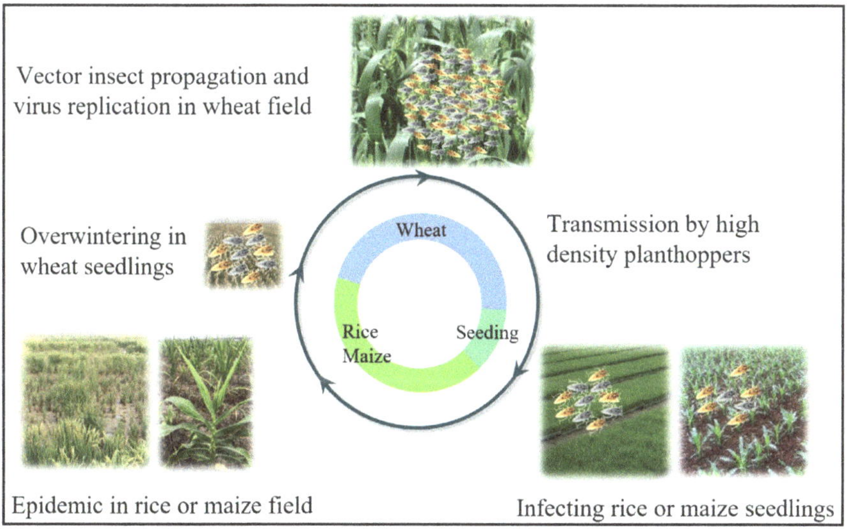

Life cycle of the leafhopper vector of rice dwarf. *Photo credit: Creative Commons.*

References

Bennett, C. W. 1967. Epidemiology of leafhopper-transmitted viruses. Ann. Rev. Phytopathol. 5: 87-108.

Fukushi, T. 1934. Studies of the dwarf disease of rice plant. Journal of the Faculty of Agriculture, Hokkaido Imperial University. 37: 41-164.

Hillman, B. I. 2001. *Phytoreoviruses*. Pages 770-771 in: Encyclopdia of Plant Pathology, vol. 2. O. C. Maloy and T. D. Murray, eds. John Wiley and sons, Inc. New York, 1346 pp.

Scholthoff, K.-B. G., 2004. *Tobacco mosaic virus*: a model system for plant biology. Ann. Rev. Phytopathol. 42: 13-34.

Slykhuis, J. T. 1976. Virus and virus-like diseases of cereal crops. Ann. Rev. Phytopathol. 14: 189-210.

CHAPTER 21

The Viroid Story

Introduction

Viroids are a novel class of pathogens and are the smallest known agents causing infectious diseases. Currently they are known to infect plants only, not animals or humans. They consist of tiny, infectious RNA molecules about 80 times smaller than the smallest known virus genome.

The discovery of the viroid is universally, and rightfully, credited to T. O. Diener, however there have been a number of other scientists, largely forgotten today, who additionally made important contributions to identifying and characterizing this curious plant pathogen. Their story, and the history of the viroid itself are the rest of the story.

A half dozen small, irregularly-shaped potatoes affected by the spindle tuber disease on the right, compared with a single healthy one on the left. *Photo credit: Nebraska Agriculture Experiment Stastion.*

Characteristics

The first viroid was discovered by T. O. Diener in attempts to purify and characterize the causal agent of the potato spindle tuber disease (PSTVd), previously thought to be caused by a virus. Like viruses, viroids invade their host's cells, forcing them to produce new copies of the viroid's RNA instead of its own. However, unlike RNA viruses, viroids have no protein coating. Furthermore, no evidence of a helper virus was ever detected in infected tissues, thus suggesting that the RNA was replicating by itself.

Theodor O. Diener

Theodor Diener is a now retired Swiss-American plant pathologist who worked from 1959 to 1989 as a charter member for the Pioneering Laboratory for Plant Virology at USDA-ARS Research Center in Beltsville, Maryland. In 1989, he accepted a position as Professor of Life Sciences at the University of Maryland's Institute of Biotechnology, from which he retired in 1999 as Distinguished University Professor Emeritus.

First Report

In 1966, Diener and colleague William Raymer, were able to demonstrate that PSTVd was not a virus using density gradient centrifugation and RNA enzyme chemistry. Diener did not discover that the potato spindle tuber disease was an infectious disease that could be transmitted from diseased plants to healthy ones. That feature of the disease had previously been determined in the 1920s by E. S. Schultz and D. A. Folsom by tuber grafting and inoculating mechanically through infected plant sap.

Instead, he disclosed that the pathogenic agent causing the disease was a free strand of RNA with a much lower molecular weight that any known virus genome, and no detectable nucleoprotein, characteristic of a virus. Diener realized these data would be subject to scrutiny and doubt, but still published the preliminary findings in 1967. He then spent the next 4 years verifying experiments and preparing to meet the skepticism he expected. His uncertainty prevented and delayed the publication of a more definitive report until 1971.

Where Did the Viroid Get Its Name?

Diener additionally realized that PSTVd was not likely the only existing agent with these characteristics, thus he wanted to affix a name for other diseases found in the future that may be similar to potato spindle tuber. In 1971, he introduced the term "viroid", meaning virus-like to signify similar pathogenic nucleic acids like PSTVd. However, he was not aware that this name had been theoretically coined independently 20 years earlier.

By 1946, the symbiotic relationship with "bacteroids" that resided in ants, termites and cockroaches was well known. They are not parasitic, but instead are required for the survival of their hosts. Due to this fact, geneticist Edgar Altenburg speculatively stated that it was also conceivable that an ultra-microscopic organism, similar to a virus, could exist universally within the cells of large organisms, serving as useful symbionts. He suggested "viroid", a name that could be analogous to bacteroids and presumably contribute some form of value to their hosts through various cell processes. It was also possible that they were organisms never noticed before by biologists, but if they should mutate, they might change, becoming pathogenic agents.

Joseph Semancik

By the early 1970s, another scientist reached a similar conclusion at roughly the same time working on another disease and plant species. Competition for this discovery was rapidly growing by other researchers as well, so Diener published the seminal report in 1971 describing and naming the pathogen "viroid". Joseph Semancik, with the University of California at Riverside, was that other competitor. He had discovered that the citrus exocortis disease was also caused by another distinct viroid (CEVd), and published it in 1972 shortly after Diener's landmark report.

Semancik's independent work was considered to be highly significant as it provided an additional, non-partisan verification of the existence and pathogenicity of these organisms from a similar but distinct disease on a different woody host. It was also said that his findings may also have greatly assisted in convincing the scientific community to accept this novel concept–thus Semancik and Diener shared the prestigious Alexander Humboldt Research Award in 1975.

Conclusions

The viroid and the potato spindle tuber disease never became a huge agricultural problem, but they were important as a model, modifying the study, diagnosis and treatment of viral plant diseases. They additionally changed the general approach and attitudes toward the study of animal diseases. Until this discovery, scientists believed that no organism could replicate itself without protein, even with the help of the host's cells. Furthermore, something this small was not supposed to be capable of infecting anything. The knowledge obtained from studying PSTVd and CEVd opened the door to the identification of additional viroids, now numbering approximately 30 species grouped into two families, Pospiviroidae, and Avsunviroidae. ***Now you know the rest of the story.***

References

Altenburg, E. The viroid theory in relation to plasmagenes, viruses,cancer, and plasmids. Am. Nat. 80: 559-567.

Centro de Astrobiologia CSIC-INTA. 2009. "Viroids: molecular vestiges of the RNA world." ScienceDaily. 17 May 2009.

Diener, T. O. (2018). Of viroids and prions. Viruses, 10(12), 663. https://doi.org/10.3390/v10120663.

Diener, T. O. 1971. Potato spindle tuber "virus" IV, A. replicating, low molecular weight RNA. Virology 45: 411-428.

Diener, T. O., and Raymer, W. B. 1967. *Potato spindle tuber* virus: a plant virus with properties of free nucleic acid. Science 158: 378-381.

Schultz, E. S., and Folsom, D. A. 1923. A spindling disease of Irish potatoes. Science 57: 149.

Semancik, J. S., and Weathers, L. G. 1972. Exocortis disease: evidence of a new species of "infectious" low molecular weight RNA in plants. Nat. New Biol. 237: 242-244.

CHAPTER 22

The Discovery of Phytoplasmas

Mycoplasmas

Mycoplasmas are small parasitic organisms that have long been known to cause various diseases in animals. Their spherical to ellipsoidal bodies are intermediate in size between bacteria and viruses, and are the smallest known prokaryotes. A prokaryote is usually a unicellular organism that contains RNA ribosomes and a double-stranded DNA genome, but lacks a cell wall or membrane-bound nuclei, mitochondria or any other organelle.

Phytoplasmas

Phytoplasmas are wall-less prokaryotes that have been associated with more than 200 "yellows-type" plant diseases over the world. These diseases were presumed to be caused by viruses due to similar symptoms (stunting, and chlorosis), and because no microbe could ever be consistently isolated from affected plants. Like viruses, the causal agents retained the ability to infect hosts after passing through bacterial-proof filters, and no known, identifiable organism could ever be correlated with the symptomatic tissues by light microscopy.

Phytoplasmas all have similar characteristics: restriction to phloem tissues, sensitive to the tetracycline antibiotics, and are transmitted by grafts, leafhoppers, and psyllids, but not by mechanical sap inoculation. However, they differ from plant pathogenic bacteria by being unculturable on synthetic media and possessing an exterior membrane instead of a cell wall.

Thus, they are not sensitive to the antibiotics, such as penicillin that act on bacteria by inhibiting cell wall formation. After their discovery, they were initially referred to as mycoplasma-like organisms (MLOs) due to similar morphological characteristics with those organisms. However, did you know that this group of pathogens was not technically discovered by a plant pathologist, but by a very attentive Japanese veterinary microbiologist? That is the rest of the story.

Discovering Phytoplasmas

The discovery of phytoplasmas was first reported in 1967 by Yoji Doi and others in the Laboratory of Plant Pathology at the University of Tokyo. At that time he was a graduate research student in the Department of Agricultural and Environmental Biology, Graduate School of Agricultural and Life Sciences.

The actual detection of this new plant pathogen may not have occurred without the serendipitous, but keen observations by a veterinary researcher named Kaoru Koshimizu. Working with electron microscopy thin sections of phloem tissues from plants affected by the mulberry dwarf disease, the plant pathology group uncovered pleomorphic, mitochondrial-sized bodies, but no virus particles were detected. Koshimizu accidentally saw some of the electron micrographs and noted the strong resemblance of the bodies with similar structures he found in cells from birds infected by a mycoplasma with which he was working. For further conformation, he also suggested the pathology group try injecting tetracycline antibiotics as a potential cure for the plants. Although this exercise did not work completely, after treatment the bodies in the phloem of infected plants temporarily disappeared with a corresponding remission of the disease, providing strong evidence for the suspected identity of this microorganism.

After this discovery was reported, the list of additional phytoplasma diseases expanded rapidly as more samples from once-thought virus diseases were re-examined by electron microscopy. It also became evident that these plant disease agents were different enough to no longer be considered true Mycoplasmas. Phylogenetic analyses of other MLOs revealed a monophyletic group not closely related to other prokaryotes. Thus in 1994, it was proposed that they become recognized under the status of the Candidatus genus *Phytoplasma* (plant-infecting mycoplasmas), representing a unique group of phloem-limited prokaryotes lacking cell walls. ***Now you know the rest of the story.***

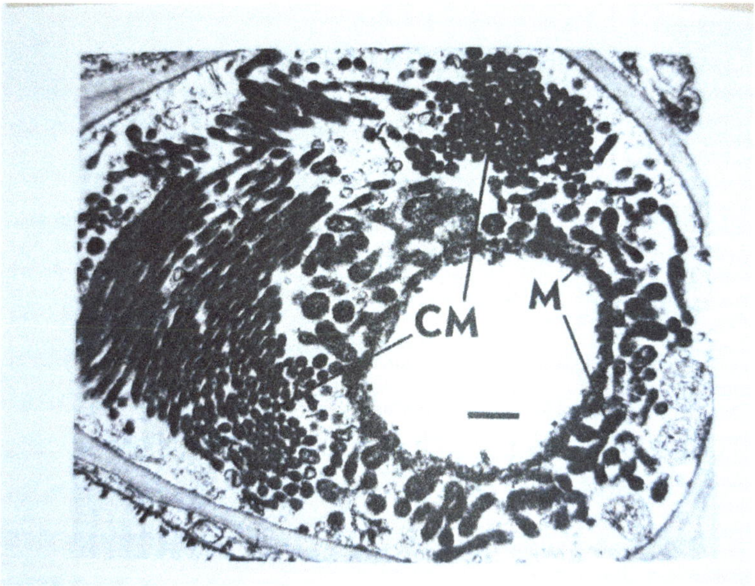

Yellows-infected peach phloem cell, packed with mycoplkasma-like bodies (M): some have aggregated n clusters (CM). *(Reproduced, by permission, from A. L. Jones, G. R. Hooper, D. A. Rosenberger, and J. Chevalier. 1974. Mycoplasmalike bodies associated with peach and periwikle exhibiting symptoms of peach yellows. Phytopathology 64:1154-1156 [Fig. 3])*

References

Doi, Y., Teranaka, M., Yora, K., and Asuyama, H. 1967. Mycoplasma- or PLT group-like microorganisms found in the phloem elements of plants infected with mulberry dwarf, potato witches' broom, aster yellows, or Paulownia witches' broom. Ann. Phytopathol. Soc. Japan 33: 259-266.

Fletcher, J. Wayadande, A. 2002. Fastidious vascular-colonizing bacteria. The Plant Health Instructor. DOI: 10.1094/PHI-I-2002-1218-02.

Harveson, R. M. 2015. The bacterium of many colors. APS Press, St Paul, MN, 288 pp.

Kirkpatrick, B. C. 2001. Phytoplasmas. Pages 769-770 in: Encyclopdia of Plant Pathology, vol 2. O. C. Maloy and T. D. Murray, eds. John Wiley and sons, Inc. New York, 1346 pp.

Maramarosch, K. 1979. How mycoplasmas and rickettsias induce plant disease. Pages 203-217 in: Plant disease, an advanced treatise, vol II. J. G. Horsfall and E. B. Cowling, eds. Academic Press, Inc. 466 pp.

CHAPTER 23

Peach Yellows

The First Phytoplasma Disease

Peach yellows is not known on any continent other than North America, suggesting that it is a native disease not noted until the highly susceptible host, the peach, was introduced. Early reports from colonists indicate that the disease had been recognized by the mid-1600s in Massachusetts. However, the first formal discovery of the disease is generally attributed to a farm in southeastern Pennsylvania in 1790.

Distribution

By 1820, the disease had spread beyond the mid-Atlantic, and by 1840 was being reported throughout the Northeastern United States and the Great Lakes. Hundreds of orchards and thousands of trees were destroyed, causing sizeable areas to abandon peach production. Seven major epidemics were documented in 1791, 1807, 1817-1821, 1845-1858, 1874-1878, 1886-1888, and 1920, occurring as far south as Virginia, west to Arkansas, northeast Texas, north to Canada and east to New England.

For years, peach yellows frustrated and puzzled scientists on its causal agent and why disease outbreaks were so sporadic but severe for a period of time, before suddenly vanishing. It can also be argued that this disease was nearly responsible for the destruction of the career of the eminent phytobacteriologist, Erwin F. Smith, before it really began. That is the rest of the story.

E. F. Smith

Erwin Frink Smith was born in Gilbert Mills, New York in 1854, but grew up in Michigan where his family moved to farm.

After graduating from high school he was employed by the Michigan State Board of Health in Lansing. One of his assigned duties was to collect and examine the literature on hygiene and epidemiology of human diseases, while also studying intermittently at the Michigan Agricultural College (now Michigan State University). This influence, combined with the supervision of two of Michigan's pioneers in medical bacteriology, Victor C. Vaughn and Frederick V. Novy, fueled an interest in this new subject.

In 1885, he transferred to the University of Michigan, where his interest in botany and plant pathology was sparked by working with the potato rot disease (late blight) during his senior year, prior to earning a bachelor's degree in 1886. The quality of Smith's work brought attention to Frank Lamson-Scribner, the first scientist commissioned by the USDA to study diseases of economic plants. Smith was hired as an assistant in the newly created Mycological Section of the Division of Botany in September, 1886.

Smith Takes on Peach Yellows

After boldly requesting the most difficult problem available for inspection, Smith was assigned to investigate a baffling peach disease known as "yellows". He spent the majority of the next seven years on this disease with extensive, meticulously performed field and greenhouse research.

Smith's early professional career was jeopardized when he refused orders to finish the studies and speculate on the disease's cause. In order to justify continued funding from Congress for the nascent department, Scribner needed definitive results on the disease and its cause. Smith's independent investigations utilized a systematic elimination approach, demonstrating the problem was not due to winter injury, insects, or soil fertility deficiencies.

Smith submitted reports in 1889 and 1894 containing valuable information consisting of comprehensive histories and discussions of environmental conditions necessary for disease development. They also included thorough descriptions of symptomatology and demonstrated that the disease was transmissible by grafting. However, Smith's results were not enough to adequately satisfy Scribner, and to his life-long disappointment, never able to identify the cause of peach yellows. This is understandable today as this pathogen was not discovered without the aid of the electron microscope.

When asked why he turned to other problems, Smith quoted: "After some years I abandoned this research and devoted my time to other subjects, mainly, as I have often said, to save my reputation, but really because the problem appeared to me to be insoluble in our then state of knowledge". He also told long-time assistant Frank Rand, "I was tired of butting my head against a stone wall". Fortunately for us he did not leave plant pathology–he went a different direction to fame.

Source of Infections

We now know that the cause of peach yellows is a leafhopper-transmitted phytoplasma, and that the native wild plum, (*Prunus americanum*), is an asymptomatic host of the pathogen and preferred host for the vector, the plum leafhopper (*Macropsis trimaculata*). This insect is found throughout eastern North America, from Quebec to Virginia, west to Ohio, Illinois, and Indiana. Although the range of the wild plum goes further south, peach yellows is present only as far as the range of the leafhopper vector.

We can now also explain the perplexing ebb and flow of yellows incidence and severity over the last 300 years. Highly disease susceptible peach trees lead to rapid mortality, removing a reservoir for pathogen survival. Furthermore, dormancy eliminates a year-round host for leafhopper vectors. Therefore, overwintering for both vector and pathogen is required in wild plums before new infections can occur the next spring when the infested leafhoppers migrate into peach orchards after dormancy is broken. Given the available resources, there is no way Smith could have succeeded at that time in history. ***Now you know the rest of the story.***

Peach yellows-infected trees, showing clusters of shoots emerging from surviving branches. *Photo credit: Hesler and Whetzel, Manual of Fruit Diseases, 1917*

References

Campbell, C. L. 1983. Erwin Frink Smith, pioneer plant pathologist. Ann. Rev. Phytopathol. 21: 21-27.

Campbell, C. L., Peterson, P. D., and Griffith, C. S. 1999. The formative years of plant pathology in the United States. APS Press, American Phytopathological Society, St. Paul, MN, 427 pp.

Harveson, R. M. 2015. The bacterium of many colors. APS Press, St Paul, MN, 288 pp.

Jones, L. R. 1939. Biographical memoir of Erwin Frink Smith (1854-1927). National Academy of Sciences Biographical Memoirs 21: Washington D.C.: National Academy of Sciences.

Rand, F. V. 1928. Erwin F. Smith. Mycologia 20: 181-186.

CHAPTER 24

Creating a Domestic Sugar Beet Seed Program in the United States

Introduction

As readers in this book may have noticed, I am fascinated with and am a passionate student of "The Big War" (WWI). In several previous chapters I have related how plant diseases influenced or affected the First World War, but this chapter will relate how it has impacted the promotion and development of a major crop industry today in the U.S.

In the early 1900s the sugar beet was unique among American crops due to its almost complete reliance upon Europe as a source for providing seed each season. At this time, the vast majority of all seed used in the U.S. was produced in Europe, primarily Germany and France. The appearance of two primary factors over the first three decades of the 20th century was instrumental in finally convincing Americans that a domestic seed industry was direly needed for this crop. The first of these factors was WWI. This is the story of the birth of the sugar beet seed industry in the U.S. and how its birth was energized by WWI. It is the rest of the story.

Harvey W. Wiley and Early Research

Harvey Wiley was chief of the USDA's Bureau of Chemistry, and as early as 1887, reported that he had successfully produced sugar beet seed; and this announcement was made prior to any commercial production in the U.S. The first serious effort to produce seeds in the U.S. was conducted by Wiley in 1891 in Schuyler Nebraska, and over the next decade he achieved steady and continuous progress from multiple sites containing diverse soil and environmental conditions stretching from Michigan to Idaho and Washington. Wiley's research not only provided evidence that production was feasible, but also that American-grown seed was of better quality, demonstrating superior germination and producing larger taproots and higher sucrose yields than those previously imported from Europe.

Based on these preliminary results, the USDA recommended the continuation of this work and to develop regionally-adapted seed for all beet-producing regions of the U.S. However, this advice was not heeded and little effort was expended toward this goal, as the foreign supplies for seed were still available and deemed satisfactory.

Bolting (flowering) of a sugar beet plant. Seeds are produced on long stalks coming out of the roots. *Photo credit: R. M. Harveson*

Problems Begin

Beginning in 1915, the quality of seed either deteriorated or became generally unavailable, as a result of the escalating war conditions in Europe. The seed which was previously used, was not well adapted for all environmental conditions in the U.S. and was highly susceptible to diseases, yet the failure to procure sufficient supplies of seed during the war years resulted in the loss of income, and investment capital. Thus, the necessity for an American sugar beet industry became apparent instead of depending on European sources. Wiley's previous investigations took on greater significance when seed availability was completely halted by the War, and sugar companies were forced to reconsider the concept of producing seeds domestically.

Home-Grown Seed Revisited

The production of seed in America was initially modeled after programs utilized in Europe. However, no attempts were made to improve the quality or create new varieties. Sugar beet, as a biennial plant, will grow vegetatively during the first year, bulking up sucrose in the roots (purpose for commercial use today) for the following year's reproductive growth. The seeds were planted and the roots were harvested in one season, stored over the winter and then replanted the following spring for seed production.

As this idea was hastily conceived and implemented, substantial losses were suffered due to high labor costs, loss of roots in storage, and resulting poor sucrose yields ranging from a few hundred to 2,000 pounds/acre with an average of 800 pounds/acre. After WWI ended, the industry returned to the importation of seeds from Europe, which should have been avoided. Unfortunately, the lessons learned during the War were soon forgotten due to lower costs and greater convenience. Approximately 15,000,000 pounds were brought into the U.S. from Europe annually between 1920 and 1933.

However, it took that second significant factor to force us to create our own regionally adapted, disease resistant seeds without depending on Europe–the curly top disease. ***Now you know the rest of the story.***

References

Coons, G. H., Owen, F. V., and Stewart, D. 1955. Improvement of the sugar beet in the United States. Advances in Agronomy 8: 89-135.

Harveson, R. M. 2015. The bacterium of many colors. APS Press, St Paul, MN, 288 pp.

Panella, L., Lewellen, R. T., and Harveson, R. M. 2009. Breeding for disease and insect resistance. Pages 3-5 in: Compendium of beet diseases and pests. R. M. Harveson, L. E. Hanson, and G. L. Hein, eds. APS Press, St. Paul, MN.

Townsend, C. O. 1917. The present status of the sugar-beet seed industry in the United States. U. S. Dept. Agr. Yearbook 1916: 399-410.

Tracy, J. E. W. 1904. Sugar-beet seed breeding. USDA Yearbook of Agriculture.

CHAPTER 25

William Orton Begins Breeding for Disease Resistance

In a recent chapter from this book, I related to you the story of E. F. Smith and his nascent studies with Fusarium wilt diseases of watermelon, cowpea, and cotton. His work with these diseases in the early 1890s brought to light the serious threat this group of pathogens posed to the production of these crops. After completing his preliminary investigations into Fusarium wilt, Smith hired a 22-year-old New Englander fresh out of college, named William A. Orton, and assigned him the onerous task of developing control measures for the "Fusarium problem". This prescient decision opened the door to the concept of genetic resistance as a means of combating the wilt diseases, and Orton's brilliant researches in this area are the rest of the story.

Cowpea plots infested with the Fusarium wilt pathogen. Note the susceptible plant plots on the left compared with those of the disease resistant cultivar "Iron" to the right. *Photo credit: USDA*

Orton Begins Plant Breeding for Disease Resistance

In the late 19th century, the wilt disease of cotton was severely threatening the economic stability of the Southern Unites States as cotton was the most valuable commodity in this region. Because Smith thought it was the most serious problem at that time, Orton ironically began with this disease in 1899, despite the fact he had never before seen a cotton plant.

Orton hypothesized that disease resistance would be among the most promising strategies after initial failures with chemical seed treatments and soil treatments of lime and fertilizers. He made selections from desirable plants that survived to harvest on Fusarium-infested soils, and obtained additional seeds from other progressive cotton growers who also had been collecting plant selections from wilt-infested fields since 1895.

Orton observed that progeny plants from the preserved seeds thrived the following season while other plants surrounding them wilted and died, demonstrating that selection for resistant plants was a practical possibility and that resistance was a heritable characteristic. He formally began a breeding program for resistance to cotton wilt in 1900 using this process of selection and testing of resulting filial generations on heavily infested land and successfully developed several new highly disease resistant cultivars of upland cotton, including 'Rivers' and 'Centerville'. Each of these introductions grew well in locations where the susceptible types previously failed and each became parents for later wilt-resistant cultivars. Within two years he had found the solution to a problem which Smith had claimed in 1898 was among the most serious in plant pathology while also saving the most important commodity in the southern U.S.

Cowpea and Watermelon Wilts

Orton next investigated the wilt disease of cowpea, but with a different approach. In this case, he tested different cultivars on "pea sick" soils (soils heavily infested with the pea pathogen) for two years. Via this method he identified a natural resistance to wilt from an existing cultivar called "Iron". More importantly, he noted that this resistance to wilt was additionally accompanied by a strong resistance to the root-knot nematode, providing one of the first examples of the developing multiple disease resistance in crops. This synergistic relationship with the two root pathogens had been noted by George F. Atkinson in Alabama a decade earlier with the nematode and cotton wilt pathogen.

For the wilt disease of watermelon, Orton first tested all available cultivars, as attempted with the cowpea studies, but found none that were resistant. He then was required to employ yet another technique-the use of exotic germplasm through hybridization. This method eventually resulted in the creation of the first wilt resistant cultivar 'Conqueror' by crossing the highly resistant African citron melon (*Citrullus lanatus*) with the susceptible watermelon cultivar 'Eden'.

Selections made from resulting progeny from 'Conqueror' maintained the desired watermelon horticultural traits while also retaining the disease resistance after being exposed to conditions of heavy infestation, thus solving the problem. 'Conqueror' was not widely used commercially, but since that time it has been employed as a parent and numerous watermelon breeding programs have successfully produced wilt resistant cultivars, which had their beginning with this original cultivar.

Orton's Influence on Modern Breeding

William A. Orton's Fusarium wilt investigations have become landmark studies for plant pathology. Although all methods utilized the concept of crop resistance, each problem was addressed and solved with a different approach (selection and progeny testing, screening existing cultivars, and hybridization with exotic germplasm). More importantly, Orton's work provided the first practical use of breeding for disease resistance, while also demonstrating that disease resistance is transmitted as an inherited trait. ***Now you know the rest of the story.***

References

Campbell, C. L., Peterson, P. D., and Griffith, C. S. 1999. The formative years of plant pathology in the United States. APS Press, American Phytopathological Society, St. Paul, MN, 427 pp.

Coons, G. H. 1928. Some aspects of the Fusarium problem. Pages 43-92 in: plant pathology and physiology in relation to man. W. B. Saunders Company, Philadephia, 207 pp.

Harveson, R. M. 1999. Evaluation of the parasitic relationship of *Melanospora* and other allied genera with *Fusarium oxysporum*. Ph.D. thesis. University of Florida, Gainesville, 129 pp.

Harveson, R. M. 2015. The bacterium of many colors. APS Press, St. Paul, MN, 288 pp.

Jones, R. L. 1931. William Allen Orton. Phytopathology 21: 1-11.

Orton, W. A. 1909. The development of farm crops resistant to disease. U. S. Dept. Agr. Yearbook 1908: 453-464.

CHAPTER 26

The Plot 30 Story

In a previous chapter of the book, I covered the beginnings of breeding for disease resistance in crops, initiated first by William Orton in 1899. His work with Fusarium wilts utilizing selection under high disease pressure, screening existing cultivars, and hybridization with exotic germplasm has become a classic in plant pathology. Around 1900, a plant pathologist in North Dakota named Henry L. Bolley employed one of these approaches for breeding using "Plot 30".

Many readers today are likely familiar with Bolley and his studies with the Fusarium wilt disease of flax. He developed disease resistant flax cultivars by selecting surviving plants grown on heavily infested soils in the same manner as Orton did with cotton wilt. However, those outside of North Dakota may not have heard of Plot 30. The account of this small piece of land in Fargo, North Dakota is the rest of the story.

Henry Luke Bolley

Henry L. Bolley was one of the first three faculty members hired at the newly formed North Dakota Agricultural College (NDAC), later becoming North Dakota State University, and one of the first plant pathologists in the U.S. He came to North Dakota in 1890 as the state's first botanist after completing the M.S. at Purdue University. Interestingly, after playing quarterback for Purdue's first football team, he also served as NDAC's first football coach, continuing in this role until 1899.

The Fusarium Problem

At the turn of the 20th century, flax was considered to be a crop capable of successful growth and production only on undisturbed prairie soils. This was due to a devastating wilt disease that followed flax cultivation, and always appearing after successive crops were grown in the same fields. This disease was well-known in Europe where it was managed with a complex rotation system involving flax cultivation once every 8 years. The earliest improvement efforts for flax production in the U.S. were carried out in North Dakota and Minnesota to specifically address this problem.

Henry Luke Bolley

Photo Credit: North Dakota State University Archives, Fargo.

As far back as 1890, Otto Lugger began studies in Minnesota on controlling flax wilt with fertilizer and fungicide seed treatments, as well as evaluating flax straw as a source of infection. Lugger's experiments were done in the greenhouse over the winter of 1890-1891, consisting of three small beds of clean, virgin soil planted with flax seeds. Bed one was a control with no additional treatment. Bed two was watered with an infusion obtained from Fusarium-infested straw soaked in warm water. Developing plants in this bed wilted and died rapidly. Bed three was watered with the same infusion used in bed two but after the infusion was boiled. The flax plants from this treatment remained as green and healthy as those in the control bed. Lugger admitted that he did not understand what he had found except that it "evidently was alive". The agronomist, J. H. Shepperd was impressed with Lugger's results with the flax wilt experiments and upon arrival in North Dakota in 1893 he formulated an idea and brought it to Bolley's attention. Shepperd agreed to devote a small plot of land (1/8 acre) to continuously grow flax each year "until something happens". Thus the story of plot 30 began in the spring of 1894.

Plot 30

Plot 30 is a small, rectangular parcel of land measuring 400 ft. x 110 ft. consisting of a finely textured soil, derived from an ancient lake bed. The plot is still in use and has been continuously cropped to flax since 1894, with the exception of 1897 (cropped to wheat) and potentially 1900 when cropping records were lost.

"Something happened" after 1895 when flax yields began declining rapidly, dropping from 12.3 bushels in 1898 to 7.4 bushels in 1899. In 1900, all plants were dead or rapidly wilting by July 4th, after which the plot was turned over to Bolley for his famous investigations. This experimental ground was eventually considered significant enough to be placed on the National Register of Historic Places in 1991.

Significance of Plot 30

Using this "flax sick" lot of land, Bolley identified the causal agent of the disease, declared it a new species, naming it *Fusarium lini*, and demonstrated that disease was not caused by a depletion of nutrients or other fertility issue. He then utilized selection breeding by collecting surviving plants and by 1908 was able to release seed for NDR 52 and NDR 73, the first wilt-resistant cultivars.

Although Orton and Bolley worked independently with no knowledge of each other, they share the distinction of being the first individuals in agricultural history to intentionally subject plants to disease epidemics in order to identify resistant selections. Orton conducted his stellar studies in South Carolina grower fields, while Bolley's landmark work was conducted in North Dakota on land known in the College records as "Rotation Plot No. 30". ***Now you know the rest of the story.***

References

Bolley, H. L. 1901. Flax wilt and flax sick soil. N. D. Agr. Exp. Sta. Bul. 50.

Dillman, A. C. 1936. Improvement in flax. Pages 745-784 in: Yearbook of agriculture, 1936.

Harveson, R. M. 2015. The bacterium of many colors. APS Press, St. Paul, MN, 288 pp.

Thompson, T. E., and Zimmer, D. E. 1975. Plot 30: proving ground for wilt resistant flax varieties. Farm Research - May-June: 28-29.

CHAPTER 27

How an Unrelated Invention Helped Develop Plant Pathology

Many of you are likely familiar with Johannes Gutenberg. He was a blacksmith who invented the printing press, introducing mechanical movable type printing into Europe in the 1440s. This brilliant invention played a key role in the development of the scientific revolution by making the production of printed books economical and more affordable to the masses. It also affected plant pathology hundreds of years later. That is the rest of the story.

In a book entitled How We Got to Now, author Steven Johnson contended that Guttenberg's invention of the printing press not only increased literacy and improved education, it also exposed a flaw common to a great deal of the population–farsightedness.This finding created a new market for the production of spectacles (eye glasses). Within 100 years of Guttenberg's invention, business for spectacle makers in Europe was booming, as more people were reading books and needing some assistance with visualizing small print.

The Jenssens and the Microscope

Sometime after 100 AD, the Romans invented and began experimenting with glass. They found that some shapes that were thick in the middle and thin on the edges would cause objects to become larger when viewed under the glass shapes, called "lenses". This word "lens" is derived from Latin for the lentil bean (Lens), referring to their common shape. It was also discovered that fires could be started after focusing the rays of the sun under these special lenses. As a result, these early lenses were also called burning glasses.

These simple glass lenses were capable only of powers ranging from around 6X to 10X and thus were more like magnifying glasses rather than our present definition of a "microscope". Around 1590, a father and son team of spectacle makers from the Netherlands, Zaccharias and Hans Janssen, found that after placing several lenses in a tube, objects at the end of the tube appeared greatly enlarged. This important discovery became the concept behind the compound microscope, and with it one could magnify objects much larger than with the simple burning glasses.

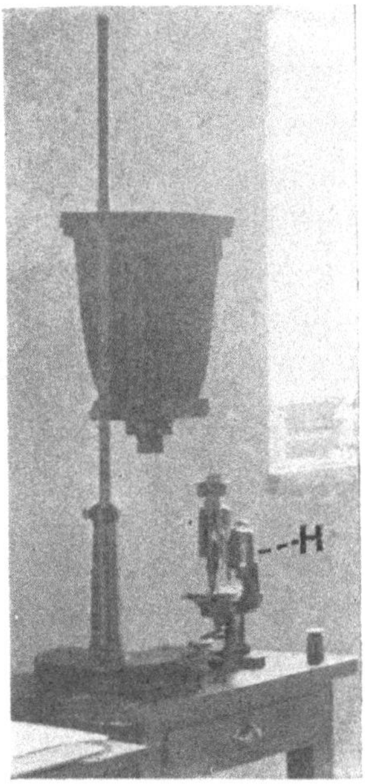

Microscope and camera system used by E. F. Smith. *Photo credit: Erwin F. Smith, USDA.*

Anton Van Leeuwenhoek

Anton van Leeuwenhoek, another Dutchman, developed new methods for grinding and polishing his own lenses. He was a draper (retailer of cloth used for clothes) who utilized the lenses to count threads in woven cloth. In 1671 he constructed his first simple microscope capable of increasing the resolution of objects up to 270X. This allowed him to view things that no one else had ever seen before, such as bacteria, protozoa, and other microorganisms, and well as yeast and blood cells. He is often referred to as the "Father of Microscopy" due to these lense improvements and his descriptions and illustrations of the small objects he was able to observe.

Microscope Allows Pathogens to be Seen

Most plant disease epidemics are caused by members of four major pathogen groups: nematodes, fungi, bacteria and viruses. In general, pathogens belonging to these groups were first recognized and studied in order of their decreasing size.

The human parasites known as Guinea worms were the first nematodes to be noticed as they were large enough to be seen by the naked eye. However, it was not until after the microscope was developed that the smaller plant parasitic nematode forms began to be observed and studied extensively.

As with nematodes, the systematic study of fungi apart from mushrooms did not formally begin until further advancements in microscopy (like those of Van Leeuwenhoek) in the 17th century. Similarly, the microscope was as influential in the discovery and characterization of bacteria as it had been with fungal and nematode pathogens in plants. Viruses were not observed until after the electron microscope was introduced in the 1930s due to their very small size.

Conclusions

Plant diseases have been known and observed by humans for millennia. The rusts, mildews, and blights were familiar to the Greeks, Romans, Hebrews, Chinese, and Indians. However, they did not understand the true nature of the diseases, and instead ascribed blame to supernatural forces, due in great part to the fact that the pathogens causing disease could not be seen.

Steven Johnson's hypothesis that the printing press spurred the need for producing magnifying glasscs for lcgions of new readers was the impetus of the idea for this article. I have taken this concept a bit further and speculated that this invention catalyzed a trickle-down response from enhanced lenses to the creation of microscopes; eventually leading to the development of plant pathology by allowing pathogens to be recognized and observed. Thus it could be argued that the printing press was indirectly responsible for the birthing of our discipline! ***Now you know the rest of the story.***

References

http://www.history-of-the-microscope.org

Harveson, R. M. 2015. The Bacterium of many colors. APS Press, St. Paul, MN, 288 pp.

Johnson, Steven. 2014. How we got to now: six innovations that made the modern world. Riverhead Books, New York, 293 pp.

CHAPTER 28

The Embarrassing Japanese Cherry Tree Incident

If you have ever been to Washington D.C. in April, you may have observed the cherry trees in bloom that line the several mile corridor stretching from the Capital building to the Presidential memorials. You may also be aware that these trees were a gift from the citizens of Japan in 1909 to be planted along the newly reclaimed Potomac waterfront. However, did you know that those flowering cherry trees that now beautify the National Mall were actually the second such donation? That fact is the rest of the story.

David Fairchild

David Fairchild, the famous plant explorer, served as the director of the USDA's Section of Seed and Plant Introduction from 1898 until 1928. He traveled all over the world collecting various plant species and bringing them back to the U.S. During his lifetime, he has been credited with overseeing the introduction of an astounding 80,000 species of plants into the U.S. One of his more high-profile accomplishments was his role in acquiring the Japanese flowering cherry trees, which he had admired from his surveys throughout Asia in the early 1900s. Fairchild worked in conjunction with an American travel photographer and writer-Eliza Ruhama Scidmore, and the wife of President William H. Taft, (Helen "Nellie" Taft) to have these cherry trees planted within the Potomac Park.

The Tafts

The First Lady was involved with the beautification of the park as a service activity and liked the idea of using these trees for this purpose. As the former Secretary of War in President Theodore

Roosevelt's administration, President Taft and his wife had previously lived in Japan, and were familiar with the appeal of the trees. Taft had been a very popular figure in Japan after his instrumental role in drafting and negotiating the peace treaty ending the Russo-Japanese War in 1905. As a result, the mayor of Tokyo chose to donate 2,000 trees as a gift from the citizens of Tokyo after learning of Mrs. Taft's interest in using them in Washington D.C..

The Trees Arrive in the U.S.

The trees from Japan arrived in Seattle in December of 1909 and were delivered to Washington D.C. by rail. Fairchild was apparently not worried about the risk of bringing in exotic pests in the name of American agriculture. However, the USDA's chief of the Bureau of Entomology, Charles Marlatt, was concerned of inadvertently importing unknown foreign pests. An inspection team was sent to examine the newly-arrived trees, including the famed nematologist Nathan A. Cobb and the first woman mycologist in the USDA, Flora Patterson.

Eradication of Infested Trees

Marlatt's report indicated that the shipment of trees from Japan had the greatest infestation of scale insects and root galls that he had ever seen. He also mentioned that Cobb had detected root-knot nematodes in 72% of the trees' root balls. Patterson additionally found that 45% of the trees were showing symptoms of crown gall on roots. Marlatt courageously recommended immediate destruction of the entire shipment via burning due to the fact that it was infested with "practically every pest imaginable". This proposition was passed all the way up the chain of command to President Taft who accepted the advice and gave the orders. A second shipment of more than 3,000 trees was donated in 1912, again in the name of the city of Tokyo "as a living symbol of friendship between the Japanese and American peoples". Prior to shipping, this new batch of trees was proactively fumigated with hydrocyanic gas.

In retrospect, the entire regrettable incident may also have served as a blessing in disguise by finally convincing Congress to pass the Plant Quarantine Act of 1912, the milestone legislation for all subsequent Federal quarantine laws. Legislation to introduce and enact laws authorizing the inspection of all plant material being brought into ports of call on a national basis had been continually attempted since 1892, but had never received enough support to pass.

Infested Japanese cherry trees being destroyed in Washington, D.C., in 1910. *Photo Credit: U.S. Department of Agriculture.*

Safeguarding the Trees

In 1914, Fairchild spearheaded the effort to propagate and distribute the trees to various American nurseries, leading to the universal planting of these cherry trees throughout the U.S. He also was instrumental in organizing and sending a matching gift of dogwood trees from the American people to the city of Tokyo.

Unfortunately, the majority of the original cherry trees (Yoshino variety) were lost due to bombing raids during WWII. In 1951 and 1980, cuttings from the original Yoshino variety sent to the U.S. were sent to Japan for restoring the Adachi Ward along the Arakawa River. In 1997, cuttings were again taken from the surviving 1912 Yoshino tree shipment and planted to insure the preservation of the genetic lineage. Additional cuttings have since been used as replacement plantings, thereby retaining the genetic heritage of the original trees sent to the U.S. in 2012. ***Now you know the rest of the story.***

Japanese cherry trees flowering in the Tidal Basin in Washington, D.C. *Photo Credit: Wikimedia Commons, Jay Wald at Wikimedia Commons.*

References

Conlon, M. 2010. Flowering cherry trees – a gift from Japan. Global Agricultural Information Network, Report number JAO507.

Lehman, P. S. 1995. Nematodes and other introduced pests that provided impetus for the establishment of regulatory agencies in the United States. Nematology Circ. No. 212, Fla. Dept. Agric. and Consumer Services, Division of Plant Industry, 4 pp.

CHAPTER 29

Was the Bordeaux Mixture Really a New Discovery?

The first new fungicides were developed after serendipitous but prudent observations. For example, brining of grains with salt water was first tried in the mid-1600s after noticing wheat seed retrieved from a sunken ship off the coast of England was free of the smut fungus. Other chemicals that were eventually tested as potential seed treatments included hot water, formaldehyde, and organic mercury products.

The elements, sulfur and copper and the heavy metals in arsenic and mercury, are all toxic to fungi, bacteria, and other organisms. They have thus played pivotal roles as components of the early fungicides developed and tested for protecting crop plants. Two of the most commonly used elements for plant disease control have been sulfur and copper.

Bordeaux Mixture

The Bordeaux (refers to a wine-producing region in France) mixture is a copper-based concoction composed of three inexpensive, easily obtainable ingredients including copper sulfate, slaked lime, and water. Its use was largely driven by severe epidemics of downy mildew in the European grape industry. Pierre Millardet is generally credited with the discovery of Bordeaux mixture in the early 1880s, despite the fact that numerous individuals had previously identified the role of copper sulfate as an effective fungicide, and had been in widespread use for more than a century. This fallacy of the fungicide's origin is the rest of the story.

Copper Sulfate

The earliest record demonstrating the fungicidal properties of copper was reported as early as 1761 by H. Schulthess with the application of bluestone (copper sulfate) as a seed treatment for smut control on cereals. In the early 1800s, Isaac-Benedict Prevost observed copper solutions killing or suppressing spore germination of the bunt fungus in the lab.

The Belgian Charles Morren promoted copper sulfate (as a soil drench) for use in controlling potato late blight during the height of the epidemic in Europe in the mid-1840s. American contemporaries Judge Cheever and James Teschemacher both advocated the use of copper sulfate as a late blight protectant. Cheever became aware of its use in France after reading a translation of Prevost's work for cereal smuts and theorizing that late blight was caused by a similar fungal pathogen.

Teschemacher was a horticulturalist and chemist who reputably learned the benefits of copper sulfate from a translation of Morren's work and prompted its use combined with lime as both a seed and soil treatment for potatoes. However, he and Morren missed their opportunity for achieving notoriety due to their emphasis on using the product as a seed and soil treatment. This mistake also delayed the discovery and use of copper-based fungicides on foliage for another forty years.

The French Wine Industry and Copper

French viniculture in the mid-1800s induced a common and inexpensive chemical that was familiar to grape-growers throughout southern France, called Verdigris, (basic copper acetate). It was a by-product after layering sheets of copper with the pressed, fermenting skins of grapes after juice extraction. One of its unique uses was combining with lime water and splattering the sickly blueish-green substance on vines late in the season near roads with the purpose of suppressing the sampling of grapes by passing pedestrians.

Copper sulfate had also routinely been used for decades as a wood preservative for the poles upon which grape vines were trained to protect them from weathering and rotting. As a result of this practice, it was observed that mildew was reduced on plants in the vicinity of the treated poles compared to those plants distant from the stakes.

The widespread usage of copper sulfate with lime made the discovery of its fungicidal characteristics inevitable. Millardet noted these relationships and theorized that the disease could be managed with the proper chemicals. He then experimented with combinations of elemental salts, table salt and lime, and found that the original mixture of copper sulfate and lime, currently being commonly used in the vineyards throughout France, worked the best. He was aware that copper sulfate had been used for more than 100 years for treating wheat seeds, but it had never been tested on foliage. The results from 1883 were promising enough to repeat the next year, but the disease incidence was low and results were inconclusive for 1884.

Pierre Marie Alexis Millardet

(Reproduced, by permission, from P. M. A. Millardet. 1885. The Discovery of Bordeaux Mixture. F. J. Schneiderhan, trans. In Phytopathological Classics, No. 3. 1933. American Phytopathological Society, St. Paul, MN)

Millardet's Dilemma

At this time, a number of other viticulture workers utilizing copper sulfate began claiming proprietary ownership for discovering the cure for downy mildew in France's vineyards. Millardet now was faced with the predicament of either keeping the results secret for another year of testing, or report on incomplete data. In order to optimize his opportunity for fame he decided to report his results. After publication, many growers in the region applied the remedy on hundreds of thousands of vines the next season with spectacular results, thereby cementing Millardet's legacy.

Both E. C. Large and James Horsfall painted a somewhat unflattering picture of Millardet and his role in the discovery of Bordeaux mixture within their historical accounts. Horsfall thought that Millardet received more accolades for this honor than was warranted, claiming that he was simply a superior, opportunistic salesman. E. C. Large was somewhat kinder by stressing that Millardet's primary contribution was recommending the usage of the fungicide with the novel purpose of applying on green foliage but hardly a "new discovery." ***Now you know the rest of the story.***

Grape plants treatment with Bordeaux Mixture
Photo Credit: Wikipedia

References

Horsfall, J. G. 1956. The fight with the fungi or the rusts and rots that rob us, the blasts and the blights that beset us. Am. J. Bot. 43: 532-536.

Johnson, G. F. 1935. The early history of copper fungicides. Agric. History 9: 67-79.

Large, E. C. 1940 and 2003. The advance of the fungi. APS Press, St. Paul MN, 488 pp.

Whetzel, H. H. 1918. An outline of the history of phytopathology. W. B Saunders Company, Philadelphia, 126 pp.

CHAPTER 30

Was the Bordeaux Mixture Really Discovered Accidentally?

As plant pathologists, we have all probably been taught that the original fungicide, copper sulfate (also known as the Bordeaux mixture) was a fortuitous accidental discovery by the French botanist and mycologist, Pierre-Marie-Alexis Millardet. The traditional narrative is that Millardet noticed healthy grapes near a roadway that had been coated with a sickly, greenish-blue substance to discourage thieving of fruit by passersby, while plants distant from the road (not treated with the unappealing material) were blighted severely by downy mildew.

Furthermore, have you ever wondered about the origin of the name for this chemical concoction–Bordeaux mixture? It has an interesting history that is likely little-known to contemporary readers. Both tales are actually somewhat mythical, and cumulatively are the rest of the story.

The use of copper sulfate (with and without lime) was common long before downy mildew reached French vineyards about 1878 and had been employed as a theft deterrent, a preservation treatment of wooden stakes for training grapevines, as well as a seed treatment for protecting cereals from smut. Millardet simply expanded the idea of using the mixture as a liquid solution applied to foliage with a hearth broom. His original recipe was prepared by combining 15 kg of lime suspended in 30 liters of water with a solution consisting of 8 kg copper sulfate dissolved in 100 liters of water.

Another Frenchman, Louis Podechard, created a recipe for a fungicide in powered form composed of copper sulfate and lime. The resulting material was then dusted on plant foliage in a method similar to that in common use for more than 30 years for applying sulfur for powdery mildew control. It is also interesting to note that the first recommendation in the U.S. for controlling potato late blight was dusting with the "Podechard powder", rather that the copper sulfate-lime mixture devised by Millardet which eventually became the gold standard for fungicides. According to George F. Johnson, the widespread knowledge and utilization of copper sulfate for multiple purposes simultaneously throughout different regions of France, made the discovery of its properties as a fungicide inevitable, and anything but "an accidental occurrence".

The apparent origin of the name "Bordeaux mixture" is also a surprising and presumably unsung story. One would assume that it was Millardet's idea based on his fame and the region of France where he lived and worked. However, in the U.S. Department of Agriculture's "Report on the Fungus Diseases of the Grape Vine" in 1886, the name of the new fungicide product was referred to as the "copper mixture of Gironde". The Gironde is a common name for the Gironde estuary, which formed at the confluence of the Garonne and Dordogne rivers before draining into the Atlantic Ocean. More commonly, it also refers to a department in southwest France, which also is the heart of the Bordeaux wine region. This publication was widely distributed by the USDA, giving full details on how to make the fungicide, but does not give it a specific name.

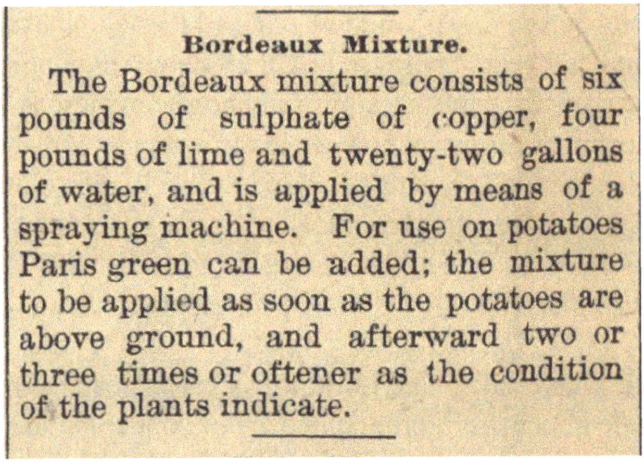

Bordeaux Mixture.

The Bordeaux mixture consists of six pounds of sulphate of copper, four pounds of lime and twenty-two gallons of water, and is applied by means of a spraying machine. For use on potatoes Paris green can be added; the mixture to be applied as soon as the potatoes are above ground, and afterward two or three times or oftener as the condition of the plants indicate.

An early recipe for Bordeaux mixture.
Photo credit: Ann Arbor Argus, June 3rd, 1892.

According to George F. Johnson, the first reference to the name "Bordeaux mixture" was found in a letter dated November 30, 1886, written to Professor C. F. Riley, USDA Entomologist by Gustave Foex with the National School of Agriculture at Montpilellier, France. Foex was relating his conclusions to Riley after attending an international congress for fungal diseases held that year in Florence, Italy. He mentioned that this formula of a copper salt was "boldly proclaimed" by the meeting attendants after hearing presentations of its virtues. Johnson further stressed that the name "Bordeaux mixture" could not have originated from Millardet as it was not mentioned as such in his first report following his extensive experiments in 1886. Johnson then suggested that the name for the fungicide must have been first coined and applied sometime between May and November 1886, presumably by members of that international congress. Regardless, the name stuck, but not really accidentally. ***Now you know the rest of the story.***

References

Horsfall, J. G. 1945. Fungicides and their action. Chronica Botanica Company, Wiltham MA, 239 pp.

Johnson, G. F. 1935. The early history of copper fungicides. Agric. History 9: 67-79.

Large, E. C. 2003. The advance of the fungi. APS Press, St. Paul MN, 488 pp.

Whetzel, H. H. 1918. An outline of the history of phytopahology. W. B Saunders Company, Philadelphia, 126 pp.

Various operations for cooking lime-sulfur wash.
Photo credit: USDA, Yearbook of Agriculture, 1906.

CHAPTER 31

Plant Diseases and Human Eating Habits

Many plant pathologists today are generally aware of the effects of some plant diseases on human society and could likely cite specific examples of how they have sometimes influenced or affected world history. From another perspective, have you ever thought about how certain historically famous diseases have driven or modified the eating habits of people? How did the cuisine of certain cultures or geographic regions originate? Examples of this concept were related to us years ago by James G. Horsfall, a very influential plant pathologist best known for his early work with organic fungicides and the creation of the Horsfall-Barrett disease severity rating scale. In my opinion, he additionally had one of the more creative and entertaining writing styles of any documenters of plant pathology history. Horsfall shared his witty thoughts more than 60 years ago on how several historic plant diseases may have altered human eating habits. His amusing narrative on this topic is the rest of the story.

Leaf symptoms of leaf rust on wheat. Notice the orange pastules.
Photo Credit: R. M. Harveson

Wheat Rust in the New World

The English immigrants settled in two primary locations (New England and Virginia) in the Western Hemisphere within roughly the same time period. Horsfall proposed that it is logical to assume that they both brought wheat with them to the New World. He also suggested that wheat rust inevitably traveled with them, yet behaved differently in the two disparate places where they established their homes.

Wheat rust is favored by a warm and wet climate. Horsfall postulated that the colonists in Virginia likely experienced greater rust problems as a result of warmer conditions than those settlers that arrived in cooler New England. The difficulty in productively growing wheat by the colonists in Virginia caused a shift to corn (maize) as a carbohydrate source, rather than the preferred wheat bread of their English ancestors. This concept is still seen today with the consumption of corn-oriented fare such as cornbread, grits or hominy being more common in the southern U.S. than in the North. He further asserted that as Americans moved west, they continued to take wheat culture with them into the Great Plains. Although in general, the southern plains (Texas and Oklahoma) were warm enough for rust, the environmental conditions of many areas were more arid than the humid southeast U.S. Therefore, being less conducive for severe rust development and infection, wheat production suffered less damage.

James Horsfall
Photo Credit: American Phytopathological Society

Cereal Consumption Habits in Europe

A similar pattern of cereal production and consumption was seen in Europe. Dietary habits were strongly influenced by differing environments and resulting plant diseases. In England, the weather was wet enough to grow wheat effectively, but too cool in general for wheat rust to be problematic. It then became the grain of choice resulting in the English favoring white wheat bread. In Italy, it is warm enough for wheat, but too dry in summers for optimal rust development, similar to the situation in the southern plains of the U.S. The Italians are well known for using wheat for both bread and pasta.

Throughout central Europe (France, Germany, Poland) wheat grows less effectively. The climate is similar to that of Virginia (warm and wet during the growing season) unlike that of England. Rust was severe in this region, thus the Germans in particular, became eaters of rye for their bread, rather than the maize of the Virginia colonists. Rye is not as susceptible to rust as is wheat. As a result of widespread cultivation of rye, ergot became hugely problematic, causing human misery and suffering throughout central Europe for centuries.

Ergot seldom occurred in Britain or the Mediterranean and thus never posed a serious problem in those locations. However, ergot began to decline in central Europe in the 18th century after the introduction of the potato. The peasant farmers rapidly discovered that potatoes provided more food per acre than either wheat or rye. Thus, potatoes greatly replaced these cereals, while also lowering the incidence of ergot in central Europe. The tremendous dependence upon the potato in Europe, particularly in Ireland, also initiated another problem–the Irish potato famine, but that is a different story for another day. ***Now you know the rest of the story.***

References

Horsfall, J. G. 1956. The fight with the fungi or the rusts and rots that rob us, the blasts and the blights that beset us. Am. J. Bot. 43: 532-536.

Horsfall, J. G., and Cowling, E. B. 1978. Some epidemics man has known. Pages 17-32 in: Plant disease, an advanced treatise, Volume II, how disease develops in populations. J. G. Horsfall and E. B. Cowling, eds. Academic Press, 436 pp.

CHAPTER 32

A Plant Disease Alters Ancient Agricultural Practices

The chickpea (*Cicer arietinum*), also known as the garbanzo bean is an annual grain legume and is a staple crop and important source of protein in central Asia, Africa, India, and the Mediterranean. It also is one the eight Neolithic founder crops responsible for the origins of agriculture, being first domesticated by early farming communities in a region known as the Fertile Crescent in the Near East, before expanding into India and Africa.

The eight founder crops include three cereals (einkorn wheat, emmer wheat, and barley); four legumes (lentil, pea, chickpea and bitter vetch); and one oil and fiber crop (flax or linseed). All of them except chickpeas follow a set pattern of fall germination, flowering in late winter-early spring, and maturation in early summer. The chickpea crop is spring-planted, and this cycle shift is thought to be due almost exclusively to a single plant disease. That amazing fact is the rest of the story.

Wild Chickpea

The wild chickpea, *C. reticulatum*, is presumed to be the progenitor of the cultivated form. It is a small, relatively rare plant endemic to southeastern Turkey and is confined mostly to shallow, limestone soils. It requires a certain vernalization period to properly germinate and yields are better in winter when adequate water levels are present. In this region most of the rainfall occurs between December and February and any crop sown after February would depend on existing soil moisture for adequate growth and flowering.

Chickpea Domestication

The chickpea is thought to have been domesticated about 11,000 years ago. The domesticated form is now planted during spring for summer harvest. It is also well known that this practice results in a substantial yield penalty. For example, chickpeas planted in the fall routinely yield 2-2.5 tons/ha more than spring-planted crops. Therefore, cropping over the winter should be the practice of choice, but it is not. Although it has been done this way for thousands of years, the question has been raised: Why did ancient farmers stop planting in the fall despite its obvious agronomic benefits? There must have been a good reason for sacrificing high yield potential for the much lower and unstable yields obtained with a system planted in spring.

This question was answered from experiments conducted in the early 1960s in Israel, and results clearly implicated a disease called Ascochyta blight as the major yield inhibiting factor. Fall-planted trials survived the winter, but were destroyed by the disease in spring. Spring sown crops were able to avoid the disease but yielded much less (<1.0 t/ha). Some fall-planted, partially disease-tolerant lines produced 70% more seed yield than the spring-planted crops, while disease-susceptible crops planted in fall yielded nothing. It was thus concluded that blight was the primary reason ancient farmers in the near East modified the planting of chickpeas to the spring and not because of freezing temperatures or some other abiotic force.

Ascochyta Blight

Asochyta blight is a fungal disease caused by *Didymella rabiei* (formally *Ascochyta rabiei*) and is considered worldwide to be the most damaging and important disease affecting chickpeas. It affects all above-ground plant organs, and under the proper environmental conditions it can be a devastating disease that spreads among and throughout fields in astonishingly short periods of time.

Conditions favoring infection include cool temperatures (65-

Ascochyta blight susceptibule chickpea variety (left) compared with the tolerant variety on the right. This shows the importance of resistant cultivars whenever possible when high disease levels are present. *Photo Credit R. M. Harveson*

70° F), high humidity, and water splashing (rainfall). Thus, disease development and spread are strongly enhanced by periods of rainfall during the cropping season, which is characteristic of the near East over late winter-early spring. A fall-planted crop would have a fully closed canopy at this time and be more susceptible to blight epidemics.

Severely affected plants wilting and dying from A. blight
Photo Credit: R. M. Harveson

Conclusions

It is now speculated that after domestication, chickpea production was abandoned as a field crop until the warm-season species were introduced from Africa and Asia before being integrated into cropping systems in the Fertile Crescent areas of the near East. Archeological records support this concept. They suggest that chickpea production began at the end of the pre-pottery Neolithic age (9,000 to 10,000 years ago), suddenly vanished, and then reappeared in the early Bronze Age (6,000 years ago) after the initiation of farming. The difficulty in growing the crop over the winter seemingly explains its scarcity during that gap period and that the conversion of planting chickpeas in spring must have started after the initiation of farming (Bronze Age). All due to a plant disease! ***Now you know the rest of the story.***

References

Abbo, S., Shtienberg, D., Lichtenzveig, J., Lev-Yadun, S., and Gopher, A. 2003. The chickpea, summer cropping, and a new model for pulse domestication in the ancient near east. Quar. Rev. Biol. 78: 435-448.

Abbo, S., Saranga, Y., Peleg, Z., Kerem, Z., Lev-Yedun, S., and Gopher, A. 2009. Reconsidering domestication of legumes versus ceraeals in the ancient Near East. Quar. Rev. Biol. 84: 29-50.

Kirst, K. K. 2016. The domestication history of chickpeas – ah! garbanzo beans! Online:https://www.thoughtco.com/the-domestication-history-of-chickpeas-170654.

CHAPTER 33

Ergot and World Historical Events

Introduction

The ascomycetous fungal pathogen *Claviceps purpurea*, causes the small grain disease called ergot. The word "ergot" is derived from the French word "argot" meaning "spur". Ergot can also refer to the sclerotia, resembling a rooster's spur, that form after infecting ovaries, displacing the developing seeds. The spur is horn-like growth or spike on the foot of a male game bird utilized for fighting.

Ergot sclerotia in a rye seed head. *Photo Credit R. M. Harveson*

Ergot has been known for millennia, but not recognized as the cause of health issues until the last 200 years. For example, an Assyrian tablet dating to 600 BC mentions noxious pustules on grain seeds, particularly rye. Documented epidemics of ergotism occurred frequently in the Middle Ages (500-1500), now known to be caused by eating ergot-contaminated rye bread resulting in the deaths of tens of thousands of people. It was most common in those colder and damp areas of Europe that were more dependent on rye. This region extended through central and eastern Europe, including from France, Germany, Czechoslovakia, Austria, Poland, and Russia–the chief areas of Europe that grew rye. It was rare in the United Kingdom where rye is not consumed as much, and the climate is less favorable for rye ergot development.

Elswhere in this book, I have addressed the supposition of ergot's role in the causation and conduction of witchcraft trials throughout Europe and colonial America. However, did you know that this disease has also been proposed as a catalyst for influencing other events in world history. These events are the rest of the story.

La Grande Peur of 1789 (the Great Fear)

The Great Fear is now considered to be a pivotal episode in the history of the French Revolution. This event was a general panic, resulting in mob mentality, violence, and riots that took place between July 17th and August 3rd, 1789. Grain yields were very poor that year with widespread shortages expected. Rumors swept through towns and villages that bandits were about to seize that year's grain harvest. In response to the rumors, fearful peasants throughout the rural provinces mobilized and took up arms, attacking their overlords' manors and looting property. Fear of the peasant revolt by the aristocracy was a determining factor in the decision to abolish feudalism.

Why this spontaneous mass hysteria occurred remains a puzzle today. Historian Mary K. Matossian claimed that one of the causes of the Great Fear was consumption of grain contaminated with the ergot fungus. In years of good harvests, grain contaminated with ergot was discarded, but when the harvest was poor, there was little other option for food. Thus, a severe ergot epidemic may have affected the mental state of French peasants, thereby contributing to the French revolution in 1789.

Demographic Depression (1750-1850)

Matossian also contended that ergot food poisoning enhanced by climate and dietary changes also played a role in population depression in certain countries, e.g. rye-consuming France compared to wheat-eating England. Fertility and mortality records showed steady growth in England with more erratic and gradual growth in France. Records also provide strong evidence that fertility suppression was greater in France during this period with decreased birth rates and increased infant mortality. As the British became less dependent upon rye, they conceived more children while suffering fewer infant deaths during the summer months of July to September.

Furthermore, during the first half of 19th century, the summers in England were not warm enough for adequate ergot development while in warmer France, conditions were better for ergot. All this resulted in higher growth rates in England, due to both fortunate weather conditions, and a dietary preference toward wheat over rye.

Effects on War

Ergot may also be responsible today for Russia's lack of possessing a warm weather seaport. In 1772, the Russian Tsar, Peter the Great tried to capture several water ports on the Black Sea from Turkey. His soldiers were stopped at Astrakhan (a city in southern Russia located on the Volga river delta where it empties into the Caspian Sea) due to an ergot epidemic. They obtained hay for their horses and bread for themselves derived from ergot-infected rye. Both soldiers and animals were affected, destroying their ability to battle the Turks, and thus were forced to retreat. The desire to expand the Russian empire has still not been fulfilled today, thus this one plant disease is now speculated to have significantly altered world history. ***Now you know the rest of the story.***

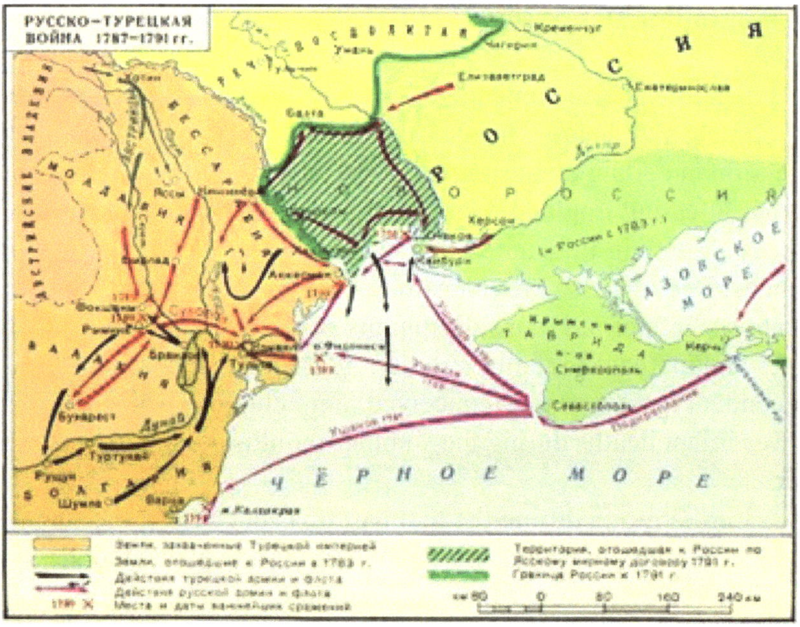

A Russian map from one of the Russo-Turk wars (1787-1791)
Photo credit: Wikipedia

References

Carefoot, G. L. and Sprott, E. R. 1967. Famine on the wind. Man's battle against plant disease. Rand McNally and Company, 231 pp.

van Dongen, P. W. J., and de Groot, Akosua, N. J. A. 1995. History of ergot alkaloids from ergotism to ergometrine. Europ. J. Obstetrics and Gynecology 60: 109-116.

Lee, M. R. 2009. The history of ergot of rye (*Claviceps purpurea*) I: from antiquity to 1900. J. Royal College Physicians of Edinburgh 39: 179-184.

Matossian, M. K. 1989. Poisons of the past: molds, epidemics, and history. Yale University Press, New Haven CN, 190 pp.

Tippo, O., and Stern, W. L. 1977. Humanistic botany. W. W. Norton and Company, New York, 605 pp.

CHAPTER 34

Ergot and the Salem Witch Trials?

Ergot is a fungal disease of small grains, particularly rye, caused by the pathogen *Clavicepts purpurea*. It enters the plant during pollination, infecting the ovaries and displacing the seeds with an "ergot" or sclerotium (a mass of fungal hyphal tissues that serve as the survival structure). The sclerotia contain several alkaloids (one of which is lysergic acid, a precursor to the hallucinogenic drug LSD) that can poison animals or humans if consumed, resulting in ergotism.

Ergot sclerotia displacing seeds within rye head
Photo Credit: R. M. Harveson

Ergotism is exhibited by two different forms (gangrenous and convulsive) and is characterized by a host of different symptoms, depending upon the specific alkaloids involved. Gangrenous ergotism causes constriction of the blood vessels resulting in poor circulation to distal parts of the body such as fingers and toes. Ultimately affected tissues die and are lost due to necrosis, and in the middle-ages this was also known as "holy fire" or "Saint Anthony's fire". Convulsive ergotism includes symptoms consistent with neurological disorders including tremors, delusions, prickling sensations on skin, convulsing seizures, hallucinations, and violent muscle spasms.

Ergot and alkaloid formation are favored by severely cold winters and a cool moist growing season. The disease also was more common in areas with high levels of rye production and consumption. Rye is more susceptible than other cereals due to flowers being open-pollinated and staying open longer thus remaining exposed for longer periods of time.

Ergot has been noted for a long time going back into antiquity. Although it has caused mass human suffering for centuries, its cause and effect was not recognized before the development of the germ theory in the mid-19th century. It has not only been responsible for many health problems, but has also been linked with dysfunctional human behavior and certain historical events. One of the more intriguing example of this condition involves its connection with the concentration of witchcraft trials in Europe and North America. That is the rest of the story.

In early modern Europe the locations of documented witch trials were clustered in central Europe (alpine France, southwest Germany), and the east coast of Scotland-all areas with cool wet weather and a dependence upon rye as a staple food. Records say that these trials decreased with the occurrence of warmer weather at the end of the 15th century. They began to increase again in the mid-1560s, with witchcraft persecution peaking in the cold century of 1560-1660.

Salem Witchcraft Theory

The worst outbreak of witch persecution in recorded history was from Colonial Massachusetts in 1692. No one has ever been able to adequately explain why it occurred so severely that year in that location, but not from other areas at other times.

One very provocative idea to explain the strange events that occurred in Salem MA in the fall of 1692 was proposed in 1976 by behavioral psychologist, Linnda Caporael. During the late summer and early fall of 1692 more than 20 people were executed after being accused, tried, and convicted for witchcraft. Much of the evidence for conviction was provided by several young teenage girls exhibiting bizarre behavior (hallucinations and sensations of being pricked or bitten), all of which are consistent with symptoms of convulsive ergotism. Furthermore, 24 of the 30 victims of the accused suffered from similar symptoms. According to English folklore, the symptoms of convulsive ergotism were also considered to be expressions of a condition known as bewitchment, resulting in the assumption and fear of witchcraft activity in Salem.

The central figure in this illustration in the courtroom is thought to be Mary Wilcott, one of the "afflicted" girls called as a witness for the Salem Witch Trials (1692-93).
Photo Credit: Wikipedia

Caporael further examined the records of weather patterns made by residents of Salem and found that cool, damp and rainy conditions had been present in spring and summer of 1692 following a severe winter. Rye was the staple grain for colonists at that time, and the grain consumed in the winter of 1691-1692 could theoretically have been contaminated by large quantities of ergot.

Thus Caporael's premise for explaining the witchcraft accusations in 1692 was an ergot epidemic after consuming contaminated rye bread, a reflection of an unrecognized health problem. This hypothesis was highly criticized by several scientists at that time but was also supported by others. Obviously, it is impossible to determine today with any certainty whether ergotism was a factor in this historical event. Nevertheless, the evidence presented, although highly speculative and circumstantial, is very compelling, and fun to ponder. ***Now you know the rest of the story.***

References

Caporael, L. R. 1976. Ergotism: the Satan loosed in Salem? Science 192: 21-26.

Christensen, C. M. 1975. Molds, mildews, and mycotoxins. University of Minnesota Press, Minneapolis, MN, 264 pp.

Matossian, M. K. 1989. Poisons of the past: molds, epidemics, and history. Yale University Press, New Haven CN, 190 pp.

CHAPTER 35

The Irish Potato Famine

Could It Have Been Avoided?

Introduction

I would assume that the majority of you readers of this chapter are at the least familiar with the Irish Potato Famine in the 1840s and its crucial role in shaping world history. The population in Ireland has been estimated to have been reduced by up to one third due to starvation and immigration during this period. Although the actual cause and effect was due to a myriad of factors, including bigotry against the Irish people, bureaucratic obstacles and poor political and governmental decisions; the famine was driven in large part by the devastating potato disease, late blight, caused by *Phytophthora infestans*.

The potato was introduced into Europe from the New World by the Spanish around 1570 and to Britain sometime between the late 1580s and the early 1590s. Its arrival into England has been commonly credited to either Sir Francis Drake or Sir Walter Raleigh's employee, Thomas Harriot. The plants found their way to Ireland, where the potato was first planted on Raleigh's estate. Although they became the staple food for the Irish people, it took some time before it was accepted. The crop was slow to be adopted due to rumors of being poisonous and being related to other well-known toxic members of the Solanaceae such as belladonna and nightshade. Furthermore, superstitions spread throughout Europe that potatoes were transmitters of leprosy and other feared diseases.

However, did you know that another New World plant was introduced into Europe about 20 years after the potato that could have displaced the potato as the starchy vegetable of choice in Europe? The narrative of this little-known crop, the Jerusalem artichoke, is the rest of the story.

Jerusalem artichoke

The Jerusalem artichoke, *Helianthus tuberosus*, also known as the sunchoke, was an important food plant for native Americans for centuries prior to the arrival of the European. The French explorer Samuel de Champlain reportedly found this plant growing at Cape Cod in 1605. The plant could reach heights of 10 feet or more, and was grown for its edible large fleshy roots. The genus name comes from the Greek words helios meaning sun and anthos meaning flower, like that of its cousin, the sunflower. The specific epithet from Latin means tuberous in reference to the edible underground tubers.

Although the roots are said to taste like the artichoke, it is not a type of artichoke, and has no connection to the city in Israel. The origin of the name "Jerusalem" is still uncertain but thought to be derived from the Italian name for the plant-*Girasole articiocco* with Jerusalem likely being a mispronunciation of girasole.

Young Jerusalem artichoke plant. Note the small edible tuber beginning to grow. *Photo Credit: R. M. Harveson*

Emerging to Europe

Tubers were brought to Europe where they rapidly attained popularity, particularly in France among the aristocracy. Gardening in French culture was a sign of prestige and nobility. Its adoption by the royalty became fashionable to everyone, thus enhancing its reputation. It quickly spread throughout the continent becoming a common vegetable food crop in a number of areas by 1650.

Sunchoke Loses Interest

In the late 17th century, the Jerusalem artichoke began to lose favor, when most of the herbalists at that time panned the plant, in large part due to its propensity to cause gastric indigestion and flatulence. Unlike potatoes, sunchoke tubers do not contain starch. They contain inulin, an insoluble sugar. Humans lack the enzymes that completely break down inulin. Thus much of it is present in undigested food when it reaches the colon, where it is converted to the digestible fructose by resident bacteria. In the process, various gasses are produced (hydrogen and carbon dioxide).

Jerusalem artichoke plants late in the season. Note their flowering while wild sunflowers are wilting and senescencing. These plants are 6-7 ft in height. *Photo Credit: R. M. Harveson*

Positive Uses

Despite these problems associated with human consumption, the plant has been used commercially in the production of fructose and as a source of industrial alcohol. The Soviets were also reputedly able to secure disease resistance in cultivated sunflowers through the use of hybrids with the Jerusalem artichoke. At one time it was even thought to potentially rival the temperate sugar beet as a perennial source of sugar after being hybridized with the common sunflower.

It was also noted that the crop was better suited to poor soils and dry conditions, easily out-yielding potatoes under these same conditions. Nevertheless, the sunchoke dropped out of site when the superstitions for the potato faded, enabling the potato to emerge as the food for the masses.

Conclusions

I have always been fascinated with virtual history and the conjecturing of "what if" scenarios, on how historical events might have been modified if certain circumstances had proceeded in different directions. In my opinion the Irish Potato Famine and the Jerusalem artichoke provide an excellent and instructive example of this concept. Had the sunchoke become the adopted staple food in Europe, historians have questioned whether the devastating potato famine in the 1840s could have been averted, thereby altering the fate and history of Ireland and subsequently, the rest of the world. ***Now you know the rest of the story.***

References

Harveson, R. M. 2015. The bacterium of many colors. APS Press, St. Paul, MN, 288 pp.

Heiser, C. B. 1976. The sunflower. University of Oklahoma Press, Norman, 198 pp.

Pappalardo, J. 2008. Sunflowers (the secret history). The unauthorized biography of the world's most beloved weed. The Overland Press, Woodstock, and New York, 256 pp.

Sauer, J. D. 1994. Historical geography of crop plants. CRC Press, Boca Raton, FL, 309 pp.

CHAPTER 36

Potato Late Blight and World War I

The elements arsenic, mercury, sulfur, and copper have played pivotal roles in the development of early fungicides as they are toxic to living organisms. Two of the most commonly used elements for plant disease control over the years have been sulfur and copper. The famous Bordeaux mixture was born after combining both ingredients, a brew composed of three inexpensive, easily obtainable ingredients-copper sulfate, slaked lime, and water.

Pierre Millardet is generally credited with the discovery of Bordeaux mixture in the early 1880s, despite the fact that copper sulfate (with and without lime) was common and had been in use long before downy mildew reached European vineyards about 1878. This fungicidal concoction had been employed for almost a century as a theft deterrent for fruit in vineyards near pedestrian walkways, a preservative treatment of wooden stakes for training grapevines, as well as a seed treatment for protecting cereals from smut. By the 1890s it was fully recognized and widely utilized as an effective method for additionally managing potato late blight.

Symptoms of late blight on a potato leaf.
Photo credit: Howard F. Schwartz and Creative Commons Attribution 3.0

Disease and World War I

Most of you are likely familiar with the infamous late blight disease of potato as a major cause of the Irish potato famine in the 1840s, and its lasting influences on world and social history. However, did you know that another late blight epidemic 70 years later played a crucial role in shaping the outcome of World War I? That is the rest of the story.

The Potato Problem Returns

In July of 1914, the bloodiest war in history to that point broke out in Europe between Serbia and Austria-Hungary. This drew Germany and Russia into the conflict, followed soon after by Great Britain and France as allies of Russia, and it would continue for the next 4 years.

The 1915 potato crop in Germany was phenomenally productive. Harvested yield was nearly three times what the German people normally needed, and more than could be consumed. This would create a large oversupply for feeding livestock. There would even be large quantities left over for conversion into alcohol–things looked very promising for the coming winter and beyond.

The major problem was a lack of sufficient storage facilities. The available warehouse space not filled with war equipment was not nearly enough to house the tremendous surplus of potatoes. The warm basements of public buildings and schools were then used for storage, but, it was soon realized that this was a mistake when tubers began to sprout and rot. Waste was immense as rotting potatoes were removed and dumped into cull piles, setting the stage for the upcoming late blight epidemic.

The Epidemic Strikes Again

The spring of 1916 was very conducive for plant growth. The new potato crop thrived from the plentiful rains and warm temperatures. Then the summer turned to cold rain. Disease began from within the masses of tubers discarded the previous year and spread rapidly throughout German potato fields aided by the cool, damp weather that summer.

There was no copper to spare for making the fungicides that could save the potato crop. Agriculture was forced to take a back seat to the military and war needs. Thus potatoes rotted in the fields without being harvested. What was harvested went to the army. The soldiers were not hungry that year, but their families back home were. By the end of 2016, German citizens were starving. An estimated 700,000 people died from starvation in the winter of 1916-1917.

The morale and desire for continuing war by the German people plummeted in 1917. After more than three years of fighting a destructive stalemate war against the Allies, they were fatigued and demoralized. Likewise, the military weakened in 1917 before finally collapsing in November 1918, ending the war.

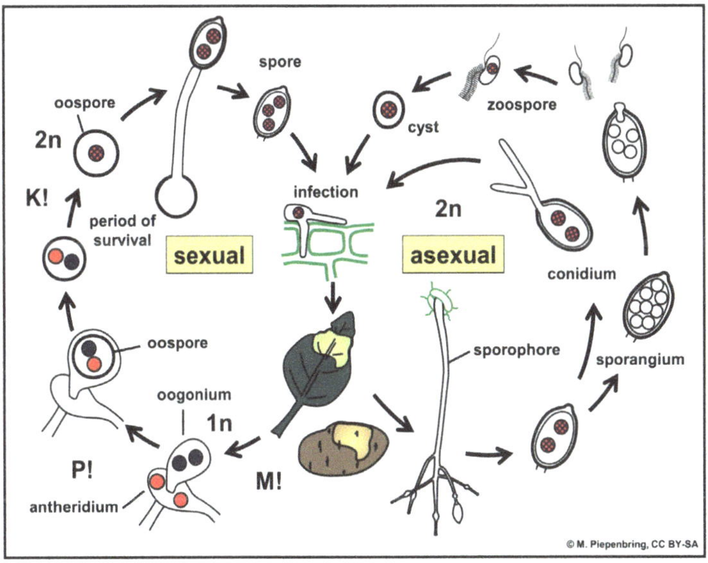

The life cycle of the pathogen causing late blight of potatoes.
Photo credit: M. Piepenbring, and Creative Commons Attribution 3.0.

Concluding Remarks

The agricultural and environmental conditions in Germany in 2015-2016 were eerily comparable to those of Ireland in the mid-1840s, with similar, catastrophic repercussions. In both situations, farmers produced a bumper crop one year with excessive quantities of unused potatoes carelessly dumped due to lack of storage facilities. The following year was characterized by a warm spring and cool wet summer, triggering widespread crop losses due to the late blight disease beginning in the cull piles, and finally culminating in famine, starvation, and economic disaster. Amazing how history repeats itself! ***Now you know the rest of the story.***

References

Carefoot, G. L. and Sprott, E. R. 1967. Famine on the wind. Man's battle against plant disease. Rand McNally and Company, 231 pp.

Horsfall, J. G., and Cowling, E. B. 1978. Some epidemics man has known. Pages 17-32 in: Plant disease, an advanced treatise, Volume II, how disease develops in populations. J. G. Horsfall and E. B. Cowling, eds. Academic Press, 436 pp.

Large, E. C. 1940. Advance of the fungi. Jonathan Cape, Ltd., 488 pp.

CHAPTER 37

Sunflower Rust and Resistance

Its Influence on Today's Sunflower Production

Introduction

In another chapter of this book, I shared with you the story of how sunflower rust served as the model that discovered the previously unknown function of the pycnial stage of rust fungi. John Craigie's landmark revelation was first demonstrated with the rust pathogen of sunflower, which proved that rust fungi had the ability to reproduce sexually, potentially creating new strains (races) of the fungi each season. This was crucial because it now explained the baffling question of how resistant cultivars that yielded well one year, would then inexplicably suffer huge losses the next from disease. However, did you know that this pathogen has also played a significant role in the creation and production of the modern commercial sunflower? That fact is the rest of the story.

The Rust Pathogen (*Puccinia helianthi*)

Rust is universally considered one of the most damaging diseases of sunflower, and is now found wherever the crop is grown. Both the sunflower plant (*Helianthus annuus.*) and the rust pathogen are native to North America, and the pathogen has followed the crop wherever it has migrated. *Puccinia helianthi* is a macrocyclic pathogen (possesses all five spore stages) and is autoecious, meaning that all spore stages occur on *Helianthus* spp.

Sunflowers Arrive in Europe

It is thought that the Spanish first introduced the sunflower from the New World into Europe through the Iberian Peninsula sometime around 1500. The plant rapidly became popular, primarily as a curiosity and an exotic ornamental plant, not as a food plant. Sunflowers spread north into England, Germany, Belgium and Holland before drifting into Eastern Europe in the 17th century. Although first utilized as an ornamental plant, seed consumption became the preferred snack in the 1740s, replacing hazelnuts. This culinary practice also resulted in gradual selection for plants with large, single heads and larger seeds. Although it was the British who first noted the plant's use as an oil crop in the early 1700s, it was the Russians who perfected this characteristic in the 1920s.

The Sunflower Migrates to Russia

Peter Romanov, also known as Peter the Great, was the Tsar of Russia from 1682-1725, and has generally, been credited with the introduction of the sunflower into Russia. After allegedly seeing the plants for the first time in Holland, he was so enamored with them that he brought the first seeds back to his homeland in the early 18th century. After the sunflower's utility for producing vegetable oil for human consumption was recognized in Russia,

Mikhail Woronin

Photo credit: Reproduced, by permission, from Phytopathological Classics #4, 1934., American Phtyopathological Society, St. Paul,MN

the crop's cultivation expanded rapidly.

Sunflower Rust and Woronin

The first serious research on the rust disease was done by Mikhail Woronin prior to his pioneering studies of the cabbage club root disease for which he is better known. During the late 1860s and early 1870s, rust became so severe that it threatened the entire industry in Russia, and Woronin was approached by the growers for assistance. He began working on the pathogen life cycle, and established the presence of both uredinial and telial spores. He furthermore conducted the first epidemiological studies, determining that the incidence and spread of disease was worsened when close row spacing was employed combined with no crop rotation. Woronin's investigations, although useful, did not solve the problem completely.

Breeding in Russia/Soviet Union

It was not until the early 1900s that intensive breeding began. Improving the sunflower with breeding was first attempted by crossing different forms of the cultivated crop with each other. The earliest record of this occurred about 1915 with the specific purpose of attaining resistance to rust, but other traits like higher yields and oil content were also pursued.

In the 1930s, the breeder and plant hunter, Nicholai Vavilov, found and collected multiple accessions of wild sunflowers in west Texas that resisted rust, increased yields, and possessed tolerance to variable environments. One of these accessions eventually served as the source of rust resistance in newly developed cultivars.

The Sunflower Returns to North America

The sunflower's homecoming has generally been credited to Russian Mennonite immigrants settling into Canada's prairie

Sunflower leaf infected with sunflower rust. Note the cinnamon-brown pustules filled with rust uredinial spores distributed on the backside of the leaf. *Photo Credit: R. M. Harveson*

provinces Manitoba and Saskatchewan as early as 1875. After its reintroduction into North America, the sunflower was used primarily as a forage and silage crop in those areas that were barely adequate for corn production, however, the rust disease continued to plague growers.

Breeding efforts for rust resistance in North America began in the mid-1930s, but were largely unsuccessful until some resistance was found in *H. argophyllis* and several wild forms of *H. annuus* from Texas. Early work at the Fargo USDA Sunflower Unit consisted of disease surveys and selecting plants exhibiting resistance to rust. These plants were then utilized in winter breeding nurseries in Hawaii and testing in North Dakota each summer. This technique resulted in a resistant line in 1973 called "Sundak", that was employed in creating the first commercial disease resistant hybrids, increasing yields up to 30%. After the risk of rust had been decreased substantially, interest, planted acreage, and production of the crop increased drastically. Fear of this pathogen has been largely responsible for improving the cultivated sunflower, which is now an important cash crop in the high plains of North America with an estimated value in 2016 approaching $470 million. ***Now you know the rest of the story.***

References

Harveson, R. M. 2015. The bacterium of many colors. APS Press, St. Paul, MN, 288 pp.

Heiser, C. B. 1976. The sunflower. University of Oklahoma Press, Norman, 198 pp.

Pappalardo, J. 2008. Sunflowers (the secret history). The unauthorized biography of the world's most beloved weed. The Overland Press, Woodstock and New York, 256 pp.

CHAPTER 38

The Relationship of Halo Blight, Kudzu, and Florence Hedges

Today halo blight is an important bacterial disease of the common bean (*Phaseolus vulgaris*), and we have been aware of it for 100 years. However, did you know that this disease in the United States could be linked with the now noxious and invasive plant Kudzu and a virtually unknown woman plant pathologist named Florence Hedges? That obscure connection is the rest of the story.

Florence Hedges
(1878-1956).
Photo credit: Wikipedia

Halo Blight

Halo blight is caused by the fluorescent bacterial pathogen, *Pseudomonas syringae* pv. *phaseolicola*. This disease first appeared in New York State in the early 1920s and was characterized and named *Phytomonas medicaginis* pv. *phaseolicola* by Walter Burkholder, a pioneering American plant pathologist. It is most destructive in areas where temperatures are cool to moderate (68 to 72° F; 17 to 20° C), and is frequently detected early in the season shortly after emergence on seedlings or during mid-vegetative stages of plant growth when environmental conditions are most likely to be within this temperature range.

Kudzu

Kudzu (*Pueraria montana* var. *lobate*) is a perennial leguminous vining plant, possessing high starch and water contents. Each plant can have a single taproot weighing up to 400 pounds, which comprises over 50% of the plant's biomass. The kudzu plant is amazingly invasive, growing up to a foot per day during warm wet summers. A single crown can produce 20 or more runners, each capable of reaching 100 feet. The plant can endure severe winters because it survives as rhizomes underground. It is now estimated to cover over 7 million acres in the United States, with the demonstrated ability to kill trees and drag down power lines.

Kudzu was first introduced into the United States from Japan at the 1876 Centennial Exposition in Philadelphia and later at the New Orleans Exposition in 1883 where it was launched to cover Japanese tea houses. It was first just a botanical novelty for inquiring gardeners. However, it later became widely marketed as an ornamental and shade plant for hot summer porches, and seeds were readily available through mail-order catalogs. The vining plant later became agriculturally important, being promoted as a high protein forage crop and as a means of reducing soil erosion in the South.

Florence Hedges, Halo Blight and Kudzu

Florence Hedges (1878-1956) was an associate pathologist with the USDA's Bureau of Plant Industry (BPI) in Washington

D.C. She began working on bean diseases for E. F. Smith after graduating from the University of Michigan in 1900, and was one of more than 20 women who worked in Smith's lab before his death in 1927.

During the same time period, halo blight on beans was being studied by Burkholder, Hedges was independently investigating a bacterial disease of the distantly related kudzu plant. In 1927, she isolated, described and named the causal agent *Bacterium pueraria*e. However, she later discovered that the bacterium causing the kudzu disease was the same as the one inciting halo blight of beans. Because Burkholder had first published the disease on beans in New York, Hedges' pathogen name (*B. puerariae*) became a synonym, before being eventually transferred to the genus *Pseudomonas*, where it resides today.

Halo blight symptom. *Photo credit: R. M. Harveson*

Nevertheless, based on Hedges' work, it was brought to light that the bacterium was potentially imported into the United States on the kudzu vine. Not only is this plant now largely and increasingly problematic as a pervasive weed pest throughout the southeastern United States, but it also is now thought to have served as a vehicle for unknowingly introducing a new pathogen into the United States that afterward caused a serious disease of the economically important common bean crop. ***Now you know the rest of the story.***

References

Burkholder, W. H. 1926. A new bacterial disease of the bean. Phytopathology 16: 915-927.

Forseth. Jr., I.N. and Innis, Anne F. 2004. "Kudzu ("*Pueraria montana*"): History, Physiology, and Ecology Combine to Make a Major Ecosystem Threat" Critical Reviews in Plant Sciences. 23: 401-413.

Harveson, R. M. 2015. The bacterium of many colors. APS Press, St Paul, MN, 288 pp.

Hedges, F. 1928. Bacterial halo spot of kudzu caused by *Bacterium puerariae* Hedges. J. Agr. Res. 36: 419-428.

Stewart, A. 2009. Wicked plants: the weed that killed Lincoln's mother and other botanical atrocities. Algonquin Books, Chapel Hill, NC, 236 pp.

CHAPTER 39

Stewart's Corny Practical Joke(s)?

Discovery

In 1895, a bacterial wilt disease of sweet corn was first discovered by F. C. Stewart, a mycologist with the New York Agricultural Experiment Station. It was causing substantial problems for sweet corn growers in market gardens of Long Island with estimated losses ranging from 20 to 100%.

Stewart obtained infected samples and from them he described disease symptoms, isolated the pathogen, and completed Koch's postulates on sweet corn using pure cultures of the bacterium. A culture was sent to the distinguished American bacteriologist, E. F. Smith who characterized and named the pathogen *Pseudomonas stewartii* (now known as *Pantoea stewartii*) in honor of the discoverer in 1898. He called the disease "the sweet corn disease of Long Island" and "Stewart's disease of sweet corn", but today it is usually termed "Stewart's bacterial wilt of corn" or simply "Stewart's wilt."

Sweet corn plots showing varieties with disease resistance (right) compared to the susceptible variety on the left.
Photo credit: USDA.

F. C. Stewart

Fred Carleton Stewart was born in the small village of French Creek, New York, but moved with his family to Iowa as a small child. He graduated from the Iowa State College with a B.S. and a M.S. in 1892 and 1894, respectively while also serving as Assistant Botanist of the Iowa Agricultural Experiment Station.

He then spent three years as mycologist for the newly created Long Island Branch Station at Jamaica, NY, where he discovered the disease that would eventually be his legacy. He furthered his education for a year with a trip to Europe, studying with several of the more prominent plant pathologists of the day. On his return in 1898, he was appointed botanist of New York Agricultural Experiment Station in Geneva, NY where he remained for the rest of his career, spanning almost 50 years.

Stewart's Practical Joke(s)?

The author of Stewart's obituary, Harry M. Fitzpatrick, described him as a modest man who shunned honors or distinction. He had an unusual personality-quiet, reserved, but with a unique sense of humor. Stewart was furthermore described as a man of conservative temperament yet "enjoyed a good joke, especially when it dealt with human interest or reaction".

Susan Gruff revealed this character trait of Stewart's in a blog pertaining to the Cornell Plant Pathology Herbarium. Gruff was the curator of the herbarium for more than 25 years, beginning the blog after retiring. In 2007, she posted the finding of an odd specimen while depositing several new ones to a pre-existing drawer labeled as containing *Stereum seashellia*, calling it "something funny in the herbarium". This blog is the rest of the story.

Upon opening the drawer, she discovered that it was filled with seashells apparently collected in Clearwater FL in 1937 and deposited by F. C. Stewart. The specimens were also labeled with a note that they were obtained from gulf beaches on the genus *Fagus*.

Herbarium drawer containing seas shells collected and accessioned as a practical joke by F. C. Stewart.
Photo credit: Cornell Plant Pathology Herbarium.

The scientific name *Stereum* refers to a genus of bracket or shelf fungi found primarily growing on trees or decomposing woody residues. She also mentions that the *Fagus* reference must refer to its identity as a genus of "beech" trees, an ingenious word play on the homophone "beach" from which the specimens were seemingly procured. Gruff also mentions that this must have been a joke perpetrated by Stewart, and it also would have required the permission of either H. M. Fitzpatrick or H. H. Whetzel in order to have been formally accessioned into the herbarium.

This story probably best demonstrates Stewart's dry sense of humor and quiet humility, obviously realizing that the eventual discovery of this hoax was likely well into the future, long after he was gone. Still, I assume, he must have privately savored this joke, particularly considering it would not be revealed for another 70 years. Based on this curious incident, I have wondered what additional surprises may be lurking, hidden covertly somewhere within the drawers in that herbarium?

I personally find this anecdotal story to be quite humorous

and admire greatly Stewart's sense of humor in conceiving and carrying out this prank on some unwitting individual in the distant future. While I was in graduate school, my doctoral co-advisor (James Kimbrough) was the curator of the University of Florida's fungal herbarium. During my tenure in Gainesville, I contributed several new species of pyrenomycetes to its collection of more than 41,000 total specimens. However, after learning of this story, I now regret the fact that I had the opportunity but not the foresight or cleverness to play a similar trick on future users of UFL's herbarium. ***Now you know the rest of the story.***

References

Fitzpatrick, Harry M. 1947. Fred Carleton Stewart: 1868-1946. Phytopathology 37: 687-697.

Gruff, S. 2007. Something funny in the herbarium. Cornell Mushroom Blog. February 5, 2007.

Harveson, R. M. 2015. The bacterium of many colors. APS Press, St Paul, MN, 288 pp.

Smith, E. F. 1920. An introduction to bacterial diseases of plants. W. B. Saunders Co., Philadelphia, 688 pp.

Stewart, F. C. 1897. A bacterial disease of sweet corn. N.Y. Agric. Exp. Sta. Bull. 130: 422-439.

CHAPTER 40

The Yellow Disease of Hyacinths

The discovery of bacteria as plant pathogens has generally been credited to Thomas Jonathan Burrill, the professor of botany at the University of Illinois with the fire blight disease in the late 1870s-early 1880s. During this same time period, another largely unknown investigator named Jan Hendrick Wakker was simultaneously studying a serious disease in the Netherlands threatening the Dutch hyacinth industry. Wakker and his early work have essentially been forgotten by history, and his unfortunate career is the rest of the story.

The Yellow Hyacinth Disease

The so-called "yellow disease" in hyacinths was first observed in 1874. By 1880 this mysterious disease had already become a serious threat to the entire industry in the Netherlands. Therefore the Bulb Growers Association (BGA) in Haarlem, Holland sought the assistance from Professor Hugo de Vries of Amsterdam University. Interestingly, de Vries was one of the scientists in the early 20th century (with German botanist and geneticist, Carl Correns, Austrian agronomist Erich von Tschermak-Seysenegg, and American wheat breeder, and economist, William Jasper Spillman) to recognize the significance of and implement Gregor Johann Mendel's work conducted a half century earlier, now referred to as Mendelian genetics or principles.

The bulb growers agreed to fund a new long-term research project to solve this problem. This task was assigned to deVries' most promising student, Jan Hendrick Wakker. Wakker spent the next winter (1881-1882) in Anton de Bary's lab learning some of the new microbiological techniques, before beginning his studies. As early as 1882, he isolated a yellow bacterium from infected bulbs, named it *Bacillus hyacinthi* (later re-named *Xanthomonas*) and published his results in 1883. He then showed the cause of the yellow disease to be this particular bacterium and followed with four additional research papers between 1884 and 1889 characterizing the etiology of the pathogen. Thus Wakker was the first to attempt proving that bacteria caused plant disease by following Koch's postulates. He published the results only seven years after Koch proved that bacteria (anthrax) caused disease in animals.

In late 1885, after Wakker had been researching this disease for 3 years, the BGA decided that the progress that had been made on the project was too little, becoming impatient with Wakker's investigations of the pathogen and its life cycle. They wanted information on disease prevention, rather than managing it after infection has taken place. They perceived that the money spent to date was not helping their plight and the fear of bankruptcy was still a viable possibility.

Wakker explained that once the bacterium entered the plant, it could not be reached, and thus the popular methods of disease control known in that day, including powders and sprays applied to external organs, would be of no benefit. He realized that some as yet unknown preventative was the only remedy that could grant their wishes.

Despite the BGA's doubts concerning his results, Wakker did determine that removing diseased plants immediately upon detection was an effective method for fighting the disease. This method was put into practice with regular scouting of fields from blossoming through harvest. Furthermore, Wakker's etiological studies proved to be useful by establishing clearly that removal of infected leaves would preserve the valuable bulb, while also demonstrating that they were the origin for bulb infection.

Wakker Leaves Plant Pathology

In 1892, Wakker left the University of Amsterdam to become the director of the Sugar Experiment Station in Java in the Dutch East Indies. Over the next five years he continued researching and publishing papers on diseases of tropical crops, particularly sugar cane. After being overlooked for three professorships (at Utrecht University and University of Amsterdam in 1896 and Leiden University in 1898), he abandoned academia forever, disappointed by what he perceived as no recognition for his abilities.

A cross section of hyacinth bulbs in advanced stages of infection from the yellow hyacinth disease. *Photo credit: E. F. Smith and USDA taken from a field near Haarlem, Netherlands, 1905.*

In fact, he was still so disgruntled in 1905 that he declined the offer for the directorship of the newly established Willie Commelin Scholten Phytopathology Laboratory, a position that was filled by the 23-year old Johanna Westerdijk. Wakker returned to the Netherlands, teaching mathematics at the Higher Burgher School (elementary) in Den Bosch, existing as a lonely and bitter bachelor until his death in 1927.

Wakker's Legacy

Although the yellow hyacinth work was Wakker's only major contribution to plant pathology, it was still accurate and faultless. The eminent American bacteriologist, E. F. Smith confirmed Wakker's results in 1897-1898 using diseased bulbs obtained from Holland. He further stated: "with the exception of some preliminary work on pear blight by Burrill, Wakker's first paper is the earliest publication of any importance on a bacterial plant disease". ***Now you know the rest of the story.***

References

Ainsworth, G. C. 1981. Introduction to the history of plant pathology. Cambridge University Press, Cambridge.

Faasse, P. A. 2008. In splendid isolation. A history of the Willie Commelin Scholten Phytopathology Laboratory, 1894-1992. History of science and scholarship in the Netherlands, vol 11. English translation by Beverly Jackson. Knaw Press, Amsterdam, 302 pp.

Smith, E. F. 1897. Wakker's hyacinth bacterium. Proc. Am. Assoc. Adv. Science 46: 274.

Smith, E. F. 1905-1914. Bacteria in relation to plant diseases, vol. 1, 285 pp., 1905; vol. 2, 368 pp., 1911, and vol. 3, 309 pp., 1914. Carnegie Institute of Washington.

Whetzel, H. H. 1918. An outline of the history of phytopathology. W. B Saunders Company, Philadelphia, 126 pp.

CHAPTER 41

Women Plant Pathologists

Early Investigations of the Dutch Elm Disease

Dutch Elm Disease

Dutch elm disease (DED) is a vascular wilt disease of numerous species of elm trees. It is caused by an ascomycetous fungus, *Ophiostoma ulmi*, and is transmitted primarily via the feeding of several bark beetle species. The disease's name originated from the fact that the early reports and studies on the disease were from the Netherlands. It was first noticed in Holland in 1919 in the aftermath of WWI. By that time, it was additionally widespread across Belgium as well as portions of northern France, killing millions of native elm trees in the process. However, only the Dutch gave it serious consideration. The early fundamental research concerning the pathogen, its vector, and the disease life cycle was conducted by Dutch scientists. Furthermore, this knowledge was obtained in large part by 7 women phytopathologists. Their little-known, ground-breaking studies laid the foundation for our current knowledge of the disease today, and serves as the rest of the story.

Johanna Westerdijk
Photo Credit: Wikipedia

Dutch Female Scientists

Barendina Gerarda Spierenburg (Institute of Phytopathology at Wageningen University) was the individual who first reported an inordinately high degree of mortality from elm trees during the 1919-1920 growing season. She also created a map showing its widespread distribution.

Johanna Westerdijk, the director of the Willie Commelin Scholten Phytopathological Laboratory led the DED investigations, and assigned specific duties to various personnel in exploring the pathogen and its various disease features. She creatively solicited research funds from every Dutch municipality through the organization of the Elm Disease Committee.

Under Westerdijk's supervision, Maria Beatrice Schwarz was the first to isolate the pathogen and prove it was the cause of DED. She named the imperfect stage, *Graphium ulmi*, as a new species to science.

Christina Johanna Buisman repeated the studies of Schwarz and confirmed that DED was caused by *G. ulmi*. She additionally identified the perfect stage of the pathogen, and discovered that two different mating types were required for sexual reproduction and formation of ascospores. She furthermore assisted an entomologist, J.J. Fransen, with his doctoral project showing that the elm bark beetles were responsible for spreading the pathogen from tree to tree. Buisman also began a program searching for resistance to the pathogen in multiple elm species, screening specimens collected from all over the world for their reactions to the disease.

Westerdijk then assigned another doctoral student, Maria Sara Johanna Ledeboer, the task of studying the physiological aspects of the pathogen. She evaluated different chemicals for toxicity to the fungus and studied its nutritional requirements for growth and reproduction. All of these activities paved the way later for injecting trees with fungicides as a therapeutic treatment.

In 1936, Johanna Catherina Went took over the resistance selection and breeding duties begun by Buisman, making the first crosses between various resistant elms, and creating new hybrids through inbreeding. Went continued this program for two decades until the mid-1950s.

Lastly, Louise Catharina Petronella Kerling was appointed the director of the Scholten Pytopthological Laboratory in 1952 with the retirement of Westerdijk, where she continued the role as director of DED research. Her primary direct contribution on this disease was a publication summarizing the disease's history, including all the contributions of the six women scientists who preceded her.

Conclusion

The impact of DED on elms in Europe and North America, although still very significant, has not proven to be as devastating as that of the similar tree disease, chestnut blight. Sanitation, rogueing, and the use of systemic fungicides and insecticide sprays have all contributed to slowing disease progress to some extent. Furthermore, even with the slow and tedious process of breeding trees for resistance, there are now resistant varieties available, thanks in great part to the pioneering work of all seven Dutch pathologists. ***Now you know the rest of the story.***

Johanna Westerdijk (center), director of the Willie CommelinScholten Phytopathology Laboratory (1917). *Photo Credit: Courtesy Universiteits Museum Utrecht*

References

Faasse, P. A. 2008. In splendid isolation. A history of the Willie Commelin Scholten Phytopathology Laboratory, 1894-1992. History of science and scholarship in the Netherlands, vol 11. English translation by Beverly Jackson. Knaw Press, Amsterdam, 302 pp.

Holmes, F. W., and Heybroek, H. M. 1990. Dutch elm disease – the early papers. Selected works of seven Dutch women phytopathologists. APS Press, St. Paul, MN, 154 pp.

Zadoks, J. C., and can Bruggen, A. H. C. 2008. Johanna Westerdijk. Pages 155-167 in: Pioneering women in plant pathology. J. B. Ristaino, ed. APS Press, St. Paul MN, 339 pp.

CHAPTER 42

Lewis David von Schweinitz

The First American Mycologist

Have any of you mycologists ever wondered about the authority of numerous fungi labeled as "Schwein" following its scientific name? Several examples include: *Puccinia helianthi* Schwein. (1822), cause of sunflower rust, and *Uncinula necator* Schwein. Burrill (1892), pathogen of grape powdery mildew.

Rust on sunflower leaves. *Photo credit R. M. Harveson*

This authority refers to Lewis David von Schweinitz, an ordained minister and bishop in the Moravian Church. He also had a passion for mycology and developed an extensive botanical herbarium with both plants and fungi he had collected, primarily from North Carolina and Pennsylvania. In the early 19th century, he was commonly recognized as the authority and founder of mycology in the U.S., and was called the first American mycologist by William Farlow, Harvard's renowned professor of cryptogamic botany. Lewis David von Schweinitz's name, life, and work in mycology are likely little known to most plant pathologists today. Thus, these facts are the rest of the story.

Schweinitz was born in Bethlehem, Pennsylvania in 1780. His parents and both sets of grandparents were early pioneers and administrators in the Moravian Church, founding the Moravian colony at Bethlehem. For example, his maternal great grandfather Nicolas Ludwig, Count von Zinzendorf established the first Moravian colony in 1722, at Herrnhut in the Görlitz district in Saxony, Germany.

His affection for mycology first began after viewing a lichen in high school. He showed such enthusiasm and aptitude for the subject, he later taught at the school prior to graduation in 1798. His father was the treasurer for the Moravian Church in America, but was called to Germany to keep books for the church. After the family's arrival, Schweinitz entered the Moravian Theological Seminary at Niesky.

L. D. von Schweinitz

Photo Credit: Reproduced from W. J. Youmans, ed. 1893-1984. The Popular Science Monthly, Vol 44. D. Appleton, New York.

It was there that he met and interacted with J. B. d'Albertini, a professor who shared his passion for fungi and inspired and encouraged him to continue mycological studies.

After graduating from the seminary in 1801, he began teaching at the Moravian school in Niesky, while also studying mycology with d'Albertini. In 1805 they jointly published a 400-page treatise on new genera and species of fungi in that region of eastern Germany. After his ordination as deacon in 1808, he was called to pastor a congregation in Gnadau. He returned to the United States in 1812, after being appointed as the administrator for the entire Moravian Southern Province of America in Salem, NC.

Schweinitz remained in Salem for most of the next decade, continuing his devotion to fungal studies and collecting botanical specimens while still maintaining his administrative duties within the church. He published seminal papers in mycology (all written in Latin, as was the practice of the day), collaborating and exchanging specimens with some of the prominent botanists at that time. His two landmark publications ("Synopsis Fungorum Carolinae Superioris" and "Synopsis Fungorum in American Borealis") established him as the authority on fungi in North America. In fact, he was so well-known, respected and admired that he was also offered the presidency of University of North Carolina, but declined.

At the end of 1817, Schweinitz traveled with his family to Germany as an elected representative to the General Synod of the Moravian Church at Herrnhut spending most of the following year in Europe. Schweinitz returned to Salem, but in 1821 was appointed as senior pastor at the Bethlehem, PA settlement where he had been born. His official duties forced him to travel extensively on church business. The trips enabled him to collect botanical specimens from new areas, but also apparently weakened his fitness and well-being. He died on February 8, 1834, after a prolonged period of fading health.

At the time of his death, his personal herbarium was the largest in the New World, consisting of 23,000 specimens, and was donated to the Academy of Natural Sciences in Philadelphia. He is now credited with the description of more than 3,000 fungal species from 246 genera, with 1,200 of those species and 7 genera being new to science. These figures, combined with another 200 new plant species, total more than 1,400 species described by this single individual-Lewis David von Schweinitz, the First American Mycologist. ***Now you know the rest of the story.***

References

Johnson, W. R. 1835. A memoir of the late Lewis David von Schweinitz, P.D. with a sketch of his scientific labors. Published by order of the Academy of Natural Sciences of Philadelphia, May 26, 1835, Philadelphia, PA.

Lewis David von Schweinitz Papers, the New York Botanical Garden: http://library.nybg.org/finding_guide/archv/schweinitz_ppf.html

"Lewis David von Schweinitz: The Botanist Bishop." The Academy of Natural Sciences. Available from http://www.ansp.org/explore/online-exhibits/stories/lewis-david-von-schweinitz-the-botanist-bishop/

CHAPTER 43

Two Early American Amateur Mycologists

In another chapter in this book, I introduced a rare rust disease of sugar beets and the extraordinary woman pathologist who first identified it, Venus Pool. I also described the pathogen's complicated life cycle and its amazingly high number of alternate hosts for the early pycnial and aecial spore stages. This chapter will focus on two largely forgotten amateur mycologists, and the expansion of known hosts for the sugar beet seedling rust pathogen, *Puccinia subnitens* in the early 20th century. Their little-known contributions and their stories are the rest of the story.

Bates and Bethel

John Mallory Bates and Ellsworth Bethel were two highly significant but little-known figures today in American mycology. Their observations contributed to our present knowledge of this rust pathogen primarily through their cooperation and collaborations with other scientists, in particular the noted rust pathologist John C. Arthur. Neither was formally trained in botany or mycology, but both were self-taught, studying fungi as a recreational hobby.

John Bates

The Reverend John Mallory Bates (1846-1930) was an Episcopalian priest, and during his day was widely known and respected as an amateur naturalist and botanist. He was particularly interested in and knowledgeable of the birds, plants and rust fungi of Nebraska. He was born in Connecticut and educated at Trinity College and Berkeley Divinity School before moving west. In

1886, he moved to North Platte, NE where he served two years as rector before relocating to Valentine, NE in the heart of the Sand-

Reverend John M. Bates
Photo credit: Nebraska State Historical Society

hills region of the state, where he stayed until 1902.

It was in this post at the age of 43 that he began studies of systematic botany as a pastime, becoming a voracious collector. His private herbarium of sedges and grasses eventually numbered more than 12,000 specimens, and was purchased by the University of Nebraska for $1,000 following his death. Although he likely is best remembered for his publication on the sedges of Nebraska, he was additionally internationally recognized as an authority on plant rust diseases and ornithology. In fact, in 1899, he was licensed by the federal government to report the activity of migratory birds in the region.

As the senior Episcopalian priest in Nebraska, Bates finally settled in Red Cloud, NE in 1903 where he spent the remaining 27 years of his life as the rector in Grace Church. He died suddenly in Hastings, NE in 1930 at the age of 84, only days after attending the joint meeting of the Nebraska and Iowa Ornithologist's Unions at

Omaha where he was on the scientific program, giving a presentation entitled "Instinct and reason in birds".

Bates' Observations with Aecial Hosts of *P. subnitens* in Nebraska

In the late 1890s, *P. subnitens* was commonly recognized on species of *Distichlis* (salt-grass) throughout North America, particularly in the western United States. A close association between *P. subnitens* infections on salt-grass and similar aecial rust infections on the sugar beet cousin, common lambsquarters (*Chenopodium album*) was also noted across the Great Plains and Western U.S.

Young aecia pustules arranged in circular patterns on the underside of a common lambsquarters leaf, one of the alternant hosts of the rust pathogen, *Puccinia subnitens*. *Photo credit: R. M. Harveson*

Prior to 1904, *C. album* was the only known aecial host for *P.*

subnitens, but in the summer of 1904, Bates suggested that rust infections he observed in Nebraska on species of *Cleome* (Beeweed), *Sophia* (tansymustard), *Lepidium* (pepperweed), *Erysimum* (wallflower), and *Salsola* (Russian thistle) were related to *P. subnitens* because they closely resembled infections on lambsquarters.

Bates communicated his observations to the noted rust pathologist, J. C. Arthur and sent him samples from *Distichlis*, and infected seedlings of *Cleome serrulate*, *C. album*, and numerous other plant species covered profusely with aecia, yellow to orange in color, all of which he thought belonged to the same rust species. Arthur pronounced the disease to be identical to *P. subnitens* and confirmed Bates' deductions by inoculating these same host plants with spores and reproducing disease.

Ellsworth Bethel

Ellsworth Bethel (1863-1925) was born in Ohio in 1863, and migrated west to Colorado in 1890. He taught high school biology in Denver, while also developing a passion for the plant and animal life in Colorado. However, it was the plant rusts that held a special interest for him, and for which he was considered a leading authority.

Bethel additionally served as the Director of the Colorado State Natural History Museum in Denver from 1906 until his death in 1925, accumulating an enormous fungal collection that was deposited into the State Museum.

Ellsworth Bethel
Photo credit: Mycological Society of America

In 1917, he began work for the USDA in the office of Forest Pathology with the specific purpose of addressing that collection.

Bethel Expands the Aecial Host Range of *P. subnitens*

Bethel repeatedly observed the apparent relatedness of *P. subnitens* distributed throughout Colorado and other regions in southwestern U.S. from many of the same aecial host species as those of Bates in Nebraska. He further verified his assumptions by growing the fungi on plants in his garden, which became an open-air laboratory for this and many other rust species.

In November, 1917, as one of his first tasks with the USDA's forest pathology office, Bethel conducted a collecting trip throughout the southwest that uncovered a wealth of new information. He found a small area covering approximately 2-3 acres near Cottonwood, Arizona in what is now the Tonto National Forest. This location contained populations of *Distichlis* that were "black with rust" and what he further described as the heaviest infections on this host that he had ever seen. He visited this area again the following spring of 1918, and was very surprised to find that nearly every species of plant in this area was covered with aecia of *P. subnitens.* Aecia collected from these common desert plants represented 37 species, and 27 genera in 15 families.

By 1934, the combined works of Bates, Bethel, Arthur and others, further expanded the known aecial host range for *P. subnitens* to more than 100 species within 50+ genera, representing 24 families. This is an amazing number of hosts for one fungus!! ***Now you know the rest of the story.***

References

Arthur, J. C. 1905. Cultures of the Uredineae in 1904. J. Mycol. 11: 50-67.

Bethel, E. 1917. *Puccinia subnitens* and its aecial hosts. Phytopathology 7: 92-94.

Bethel, E. 1919. *Puccinia subnitens* and its aecial hosts II. Phytopathology 9: 193-201.

Bohlke, L. B. 1983. Grace Church, Red Cloud-a "true story of the midwest Nebraska Historic. 64: 492-511

Harveson, R. M. 2015. The bacterium of many colors. APS Press, St Paul, MN, 288 pp.

Szasz, F. M. 1988. The protestant clergy in the great plains and mountain west, 1865-1915. University of New Mexico Press, Albuquerque, 288 pp.

CHAPTER 44

Foreshadowing Common Relationships with Venus W. Pool

Early Woman Plant Pathologist

Over the last decade, I have researched and written dozens of articles on the history of agriculture in Nebraska, specifically plant pathology. As a result of this exercise, I have learned of and become particularly intrigued with the story of Venus W. Pool, and her work in the early 20th century. Her career was short but very notable as one of the first women plant pathologists in the U.S. These efforts have also uncovered several personal incidences where our academic paths have crossed, although these parallelisms are separated by a century. Pool's story and the surprisingly comparable relationships between our respective works are the rest of the story.

Venus W. Pool and Her Accomplishments

Venus Worrell Pool was born in Table Rock, NE on May 8, 1882. She attended the University of Nebraska, receiving the B.S. in 1904, and was awarded the M. A. in botany in 1908 with her thesis entitled "Tomato Fruit Rots". She was the first assistant in plant pathology on the Nebraska Experiment Station staff serving in this role from 1906 to 1911. As a research assistant she co-authored several seminal experiment station publications with both department chairs over this six-year period-Frederick D. Heald and E. Mead Wilcox. She was also an accomplished artist, contributing beautiful line drawings of fungal structures and spores included in the publications.

One of these early papers she authored with Heald (1909) was unusual and entitled: "The influence of chemical stimulation upon the production of perithecia of *Melanospora pampeana*". It was the first report demonstrating the enhancement of growth and sporulation of one fungus by another. They discovered that *Melanospora pampeana* (syn. *M. zamiae*) yielded perithecia abundantly within a mixed culture with *Fusarium moniliforme*, but only produced sparse mycelial growth in pure culture.

After special legislative funds became available, a Potato Disease Laboratory was established in western Nebraska (Alliance) to study storage diseases of potatoes. Pool was assigned to lead this effort in the summer of 1909, concentrating on a serious disease called dry rot caused by several species of *Fusarium*. George K. K. Link continued research on this topic, becoming his M. A. thesis in 1912. This work resulted in the first research bulletin (#1) of the Nebraska Experiment Station, authored by Wilcox, Link, and Pool.

In 1911, Pool left Nebraska and began with the Bureau of Plant Industry (BPI) in Washington D.C. She was hired as assistant pathologist ($1,620/year) with the Cotton and Truck Crop Disease and Sugar Plant Investigations unit stationed at Rocky Ford, CO. Impressively, she was the first woman to be given complete charge and responsibility for field experimental work in the BPI. In this position, she published several papers on foliar diseases of sugar beets, with her assistant, Marion Bertice McKay (later her husband) concerning Cercospora and Phoma leaf spots.

Another foliar disease she encountered was the heteroecious (requiring two distinct plant hosts) sugar beet seedling disease, caused by the macrocyclic (containing all spore stages) rust pathogen, *Puccinia subnitens*. In 1914, she and McKay published the first report of this disease on an economic crop (sugar beet), where it was found occurring widespread over a two-year period (1912-1913) in Colorado. They furthermore elucidated its unique but complicated life cycle and identified inland saltgrass (*Distichlis spicata*) as the primary, alternate host for its uredinial and telial spore stages. This curious fungus had been previously recognized by mycologists because of its enormous host range for the pycnial and aecial spore stages, but never as a pathogen of an economic crop.

Pool married McKay in 1915 and they relocated to Oregon where he served as a research assistant in botany with Oregon State College's Agricultural Experiment Station in Corvallis, OR. They had two children: Muriel, born in 1918 and Maurice in 1922. Sadly, Pool must have retired at that point as I could find no evidence of further research from her after 1916.

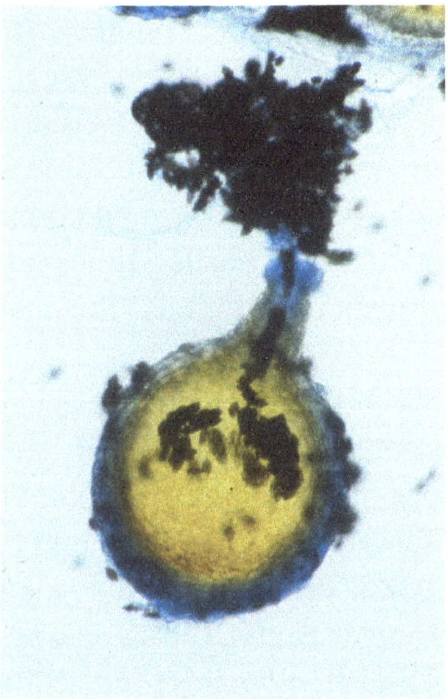

Perithecium of *Melonaspora zamiae* exuding ascospores, which can germinate and continue infecting new hosts.
Photo credit: R. M. Harveson

Coincidentally Similar Relationships

The first of these personal relationships I discovered with her work began in early 1999. I was a graduate student in Florida and accidentally discovered the novel report published by F. D. Heald and V. W. Pool in 1909 pertaining to the stimulation of growth and sporulation of one fungus by another. I referenced this paper in my doctoral dissertation because they both involved related topics, but without understanding its provenance or context. My project identified several pyrenomycetous fungi (*Melanospora zamiae*, *Sphaerodes retispora* var. *retispora*, and *Persiciospora moreaui*) and characterized their parasitic relationships with the *Fusarium oxysporum* pathogens causing wilt diseases in tomatoes and watermelons. I investigated these fungi as potential biological control agents after noting and documenting how they grew and reproduced significantly better in the presence of their *Fusarium* hosts, which Heald and Pool had observed with *M. zamiae* and *F. moniliforme* one hundred years previously.

The second incidental connection with my work involved the sugar beet seedling disease ten years later. Pool's report in 1914 provided a reference, aiding us to identify, and then publish new reports on the outbreaks of this extremely rare disease of sugar beets during the 2009-2010 seasons in Nebraska. They were the first reports for the natural occurrence of this disease in Nebraska and the first from beet production fields since the original communication by Pool and McKay almost 100 years earlier. It has not been seen since, and we now humorously refer to it as "the century rust disease".

To me, it is remarkable that both of these serendipitous events with her have occurred at two different stages in my career: one as a graduate student involving the same obscure, forgotten plant pathologist (Pool), the same obscure fungus (*Melanospora*) and both associated with *Fusarium* spp., while the other occurred as a professor affiliated with the same crop (sugar beets), the same obscure pathologist and rust pathogen (Pool and *P. subnitens*) and affiliated with the same institution, the University of Nebraska. What are the chances? ***Now you know the rest of the story.***

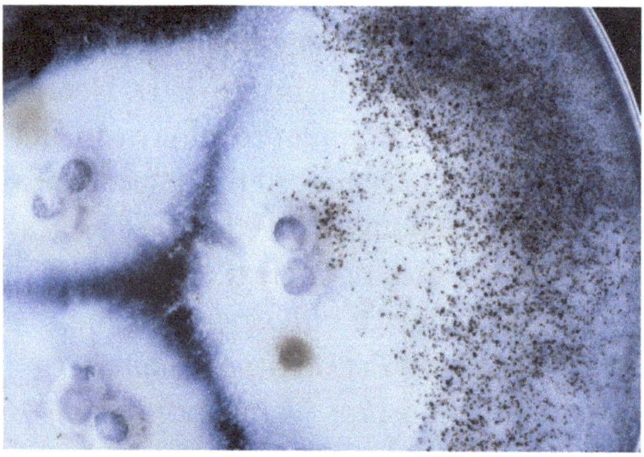

Close up image showing small black specks (perithecia containing ascospores) of *Melonaspora zamiae* within the white mycelial cultures of *Fusarium oxysporum*. *Photo credit: R. M. Harveson*

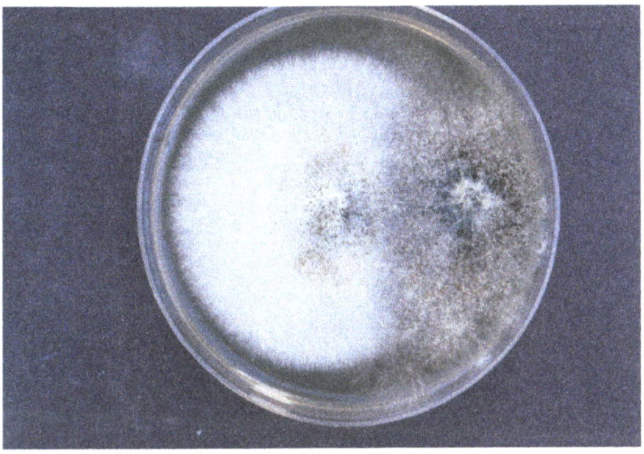

Dual culture assay showing the mycoparasite, *Melonaspora zamiae* (right) inhibiting advancement of growth of watermelon wilt pathogen, *F. oxysporum* f. sp. *niveum* (white mycelium on the left).
Photo credit: R. M. Harveson

References

Goss, R. W. 1964. The development of the plant pathology department in the Nebraska Agricultural Experiment Station. (Unpublished departmental report on file in the department of plant pathology) IANR, UNL, Lincoln.

Harveson, R. M. 1999. Evaluation of the parasitic relationship with *Melanospora* and other allied genera with *Fusarium oxysporum*. Ph.D. thesis. University of Florida, Gainesville, FL.

Harveson, R. M. 2010. First report of sugar beet seedling rust caused by *Puccinia subnitens* in Nebraska. Online. Plant Health Progress. doi:10.1094/PHP-2010-0315-03-BR.

Harveson, R. M. 2020. A century of plant pathology in Nebraska. Zea Books, Lincoln NE, 114 pp.

Heald, F. D., and Poole, V. W. 1909. The influence of chemical stimulation upon the production of perithecia of *Melanospora pampeana* Speg. Nebr. Agr. Exp. Sta. Rept. 22: 130-134.

Pool, V. W., and McKay, M. B. 1914. *Puccinia subnitens* on the sugar beet. Phytopathology 4: 204-206.

CHAPTER 45

Another Wonder Drug

Benjamin Minge Duggar was a highly influential, but today largely unknown botanist, constantly driven to learn and teach. He was born in the village of Gallion, AL in 1872, the fourth of six sons. The precocious boy entered the University of Alabama at age 14, but after developing an interest in agriculture, transferred to the Mississippi Agricultural and Mechanical College (now Mississippi State University), where he graduated in 1891 with a B.S. at the age of 19. He received the M.S. degree from Alabama Polytechnic Institute (now Auburn University) a year later. Always yearning for knowledge, he obtained additional degrees from Harvard (B.A., 1894; M.A., 1895), and Cornell (Ph.D., 1898), and made several extended trips to Europe to further his studies with several of the prominent botanists of the day.

Benjamin Minge Duggar
Photo credit: University of Wisconsin Libraries

Duggar taught at several prestigious institutions including Harvard, Radcliffe, and Cornell, and he served with the U.S. Department of Agriculture and with the State of Illinois as assistant state botanist. His love for teaching was further demonstrated by authoring three landmark, fundamental botany textbooks, "General Botany", "Plant Physiology", and in 1909 the first American textbook, (and the first written in the English language), devoted exclusively to plant pathology-"Fungous Diseases of Plants", of which I proudly possess a battered first edition copy.

Duggar's achievements in the fields of botany, plant pathology and plant physiology were eclectic, but very prolific. He produced one of the first major comprehensive treatises on the fungal genus *Rhizoctonia* in 1915, discovered the conidial stage of the cotton root rot pathogen, moving the species from *Ozonium* to *Phymatotrichum*, as well as pioneering the culture of mushrooms artificially. He was highly regarded by peers, being on committees founding the American Society of Agronomy (1907) and American Phytopathological Society (1908) as well as serving as the president of numerous additional botanical societies.

He also held academic positions in plant physiology at Cornell, University of Missouri, and Washington University (St. Louis) where he also was director of the Missouri Botanical Garden before finally settling at the University of Wisconsin in 1927. His interests turned to botanical physiology, including red pigment development in tomatoes, enzymes of the red alga *Fucus vesculosus*, and nitrogen fixation in legumes while at Missouri. At Wisconsin he focused on more basic research, conducting some of the first investigations on the *tobacco mosaic virus*, the physiological and biological effects of radiation on fungi, and quantum relations in the photosynthesis of the alga *Chorella*.

In 1943, he retired as Emeritus Professor from the University of Wisconsin at age 71. Despite the impressive list of accomplishments described above, perhaps his greatest contribution to science and humanity was yet to come. That is the rest of the story.

After retirement, Duggar turned down several lucrative consultant offers, but a chance reading of an article on his career in a scholarly journal jolted him out of his brief retirement. The article stated: "Dr. Duggar will best be remembered for his noteworthy inquiry into the physiology of mushrooms". He interpreted this comment as the description of an insignificant career, and he vowed to be remembered for more than work with mushrooms.

Therefore, he accepted a consulting position with Lederle Laboratories in 1944. At this time, new "wonder drugs" like penicillin and streptomycin were widely being employed to combat bacterial infections, but they were not effective against all bacteria. Duggar focused on soil-inhabiting organisms in the effort to identify another antibiotic product similar to streptomycin (produced by the filamentous bacterium *Streptomyces griseus*), a discovery made in 1940 at Rutgers University by the soil scientist, Selman Waksman.

Remarkably, Duggar was able to recruit soldiers and sailors, in the midst of the largest war in history, to collect soil samples from numerous sites worldwide–Europe, middle Asia, North Africa and South America. Duggar's group tested more than 3,500 soil samples before identifying an organism in 1945 that produced a gold-colored pigment. This substance exhibited antibiotic properties against an enormously wide range of bacteria–more than either penicillium or streptomycin. Duggar named his organism, *Streptomyces aureofaciens* (gold maker), a relative of Waksman's *S. griseus*. Curiously, despite testing samples from all over the world, his bacterium was found closer to home. It originated from Plot 23 of Sanborn Field, the agricultural testing station of the University of Missouri.

After 3 years of continued testing, it was concluded that the golden antibiotic chemical, christened aureomycin, was active against 90% of pathogenic bacteria, and against pathogens that did not respond to other antibiotics such as chlamydias, rickettsias, mycoplasmas, and even several viruses. Aureomycin was introduced in 1948 as the first broad spectrum antibiotic, which then triggered development of a new era of antibiotics for use in the treatment of human diseases. Duggar's legacy has definitely surpassed mushroom cultivation! ***Now you know the rest of the story.***

References

Gale, T. 1997. "Aureomycin." Medical Discoveries, Encyclopedia.com. 1 Sep. 2018 http://www.encyclopedia.com

Rosen, W. 2017. Miracle cure, the creation of antibiotics and the birth of modern medicine. Viking Press, New York, 358 pp.

Walker, J. C. 1958. Benjamin Minge Duggar (1872-1956). Biographical memoir, National Academy of Science, Washington D.C.

Walker, J. C. 1982. Pioneer leaders in plant pathology: Benjamin Minge Duggar. Ann. Rev. Phytopathol. 20: 33-39.

CHAPTER 46

The Other Smith Controversy

The Fischer-Smith Controversy

Most plant pathologists today are generally aware of the Fischer-Smith Controversy that occurred at the turn of the 20th century. Despite increasing numbers of reports concerning the implication of bacteria as plant pathogens, the European plant scientific community disputed this hypothesis and refused to accept the reports as valid. This controversy was exemplified with a series of polemic and controversial correspondences exchanged between the well-known German scientist, Alfred Fischer and the American phytobacteriologist, Erwin F. Smith.

Five articles/editorials were published in the Zentralblatt fur Bakteriologie (now known as the International Journal of Medical Microbiology) between 1897 and 1901. Two were penned by Fischer challenging Smith's accuracy and reliability as a scientist, and three from Smith stressing Fischer's lack of understanding and knowledge of bacteriology. Nothing was ever proven or concluded as a result of these communications because they did not introduce new information. They simply consisted of summarizations from previously published papers. Smith was eventually proven correct and was vindicated from Fischer's venomous responses. However, did you know that another, heated argument involving a different Smith and a plant disease occurred 25 years prior to the Fischer-Smith exchange in Europe between two individuals? That lesser-known controversy is the rest of the story.

Anton de Bary

Heinrich Anton de Bary was initially trained as a medical doctor but became interested in botany while still in medical school. In 1853, at the age of 22, he published a book on the rusts, smuts, and white rust diseases of several crops. He made important observations on corn smut concerning the development of fungal mycelium, spore-forming hyphae, and chlamydospores as well as pioneering investigations into the downy mildews and discovering heteroecism, needing two unrelated host plants in order to complete the life cycle in many rust fungi.

Anton de Bary
Photo Credit: American Phytopathological Society

De Bary became the most influential figure in mycology and plant pathology of the nineteenth century and is considered today by many to be the "Father of Plant Pathology". In 1861, de Bary published his classic work on the late blight disease of potatoes and presented conclusive evidence that it was caused by a fungus-like pathogen (oomycete). He provided an accurate description of the pathogen's life cycle and development of disease, including the formation, discharge, and germination of zoospores. He also documented the penetration and development of the pathogen within plant tissues after infection, and how it overwintered as mycelium in tubers. Because of the unique and distinctive mode of spore formation of the late blight pathogen, he created a new genus (*Phytophthora*) specifically for this disease, with *P. infestans* becoming the type species.

The de Bary-Smith Controversy

By the 1870s, the potato late blight was readily accepted universally as being induced by de Bary's newly named pathogen, but numerous other aspects of its biology and life cycle were still largely unknown. At this time, an ongoing feud concerning the mechanism of pathogen survival began between de Bary and the British illustrator, Worthington G. Smith. Although Smith was trained as an architect, he gave that up and began a second career as a freelance illustrator and mycology enthusiast. In 1875, he reported that the spores of the late blight pathogen were abundantly produced in infected potato leaves after being submerged in water. He received many laurels and a great deal of recognition for this finding, including the Knightsian Gold Medal, awarded by the Royal Horticultural Society.

Shortly after this incident in 1876, de Bary, now recognized as the authority on the late blight disease, presented a report (ironically commissioned by the Royal Agricultural Society the previous year), where he proposed his idea for the new genus designation, *Phytophthora*. His report thoroughly debunked Smith's claim by proving that the overwintering spores in question belonged to a species of *Pythium*, rather than the late blight pathogen. Nevertheless, Smith continued to espouse his thesis and was unwavering in his refusal to accept de Bary's results. He was appointed to several governmental commissions and wrote a textbook in 1884 on plant diseases, still insisting that his results were correct. It was not until 1911 that de Bary was finally experimentally acquitted after G. P. Clinton described the production of *P. infestans* oospores on oatmeal culture. In contrast to the Fischer-Smith controversy, the "Smith" in the de Bary-Smith dispute was wrong! ***Now you know the rest of the story.***

Worthington G. Smith
Photo Credited: Wikipedia

References

Ainsworth, G. C. 1981. Introduction to the history of plant pathology. Cambridge University Press, Cambridge.

Campbell, C. L., ed. and trans. 1981. The Fischer – Smith controversy: Are there bacterial diseases of plants? Phytopathological Classics 13, St. Paul, Minn. APS Press.

CHAPTER 47

The Unknown Plant Pathologist

By studying this and other chapters in this book, a reader should recognize my great interest in researching, reviving, and bringing to light little known or forgotten stories of events or people that have somehow influenced plant pathology or its development. This chapter focuses on an obscure French physician and parasitologist, Casimir-Joseph Davaine. In my opinion, he was one of the most remarkable, yet little known scientists of the 19th century, making noteworthy contributions to medical and veterinary pathology, as well as plant pathology.

Anthrax

In 1850, he and fellow pathologist, Pierre François Oliver Rayer, first observed and reported the presence of the anthrax bacterium from the blood of diseased sheep. In 1863, Davaine demonstrated experimentally that the pathogen and resulting disease could be directly transmitted from one animal to another, but did not fully recognize the implications of these results. However, Davaine's investigations occurred a full 15 years before Robert Koch proved the cause and effect of this relationship by experimentation with anthrax (Koch's postulates).

Enter Davaine

Many historians have generally attributed the discovery of bacteria as plant pathogens to the American, Thomas J. Burrill. I have also written a chapter in this book-"The Yellow Disease of Hyacinths" claiming that the Dutch plant pathologist, J. H. Wakker, should also, at the least, receive equal credit with Burrill for this revelation. However, I have since found another report by R.

H. Estey making a stronger case for placing that this honor on the work of Davaine. Estey expressed that Davaine's early work with plant diseases is as significant to phytopathology as the so-called "classical works", and the account of this scientist and his efforts is the rest of the story.

Casimir Joseph Davaine
Photo credit: Wikimedia Commons

Davaine's Phytobacterial Investigations

Estey contended that Devaine was the first researcher to successfully produce bacterial disease in plants with inoculations. In 1868, Devaine describes the appearance of a tender, moist rot that occurs on tissues of vegetables or succulent plants, followed by the development of bacteria that differ from those of anthrax. He further noted that the disease could be easily transmitted from one plant to another by inoculation, stating: "around the inoculation point an oil-like spot appears that grows and takes over the whole plant if the diseased part is not cut off". This is significant historically because it was accomplished a full 10+ years prior to Burrill's work.

Davaine was also the first to show how a plant disease could be cured. He was familiar with Pasteur's method of removing microorganisms from wine with heat (pasteurization), so he exposed infected plants to temperatures exceeding 52° C (122° F). Devaine remarked "bacteria ceased their movement, the impairment which they caused ceases to make progress, the disorganized part dries up, and the plant continues to grow as if it had never been infected". Therefore, he was the first to demonstrate heat therapy for a plant disease in 1868, almost 20 years before J. L. Jensen's famed hot water treatments for controlling smut on cereal seeds.

Nematological Investigations

In addition to bacterial studies, Devaine studied several other plant diseases including a disease of the house plant, impatiens. He also observed that fruit resisted fungal infections if the skins remained unbroken. He also noticed that some fungi induced rotting faster than others.

However, his most extensive contributions may have been in the nascent discipline of nematology. His work on the wheat gall nematode, published in 1857, greatly improved our understanding of the biology and life cycle of this nematode by providing complete descriptions of developmental growth stages of the pathogen during infection of wheat heads. He saw eggs in the bodies of females and their process of hatching, and learned that larvae could survive for many months at low temperatures as low as -20° C (-4° F) and resisted chemicals and environmental extremes better than the adult nematodes. Furthermore, his research in this area contributed to silencing the defenders of spontaneous generation, which was widely believed at the time.

Conclusions

I find the story of Casimir-Joseph Davaine and his investigations to be engaging as his observations were well ahead of their time. It is also remarkable to me that Devaine's accomplishments were all made while still maintaining a medical practice. In fact, he conducted his anthrax studies despite having no formal laboratory resources for research. His experimental animals were actually kept in the garden of a friend. Although historically, he has been better known and remembered for work with anthrax and septicemia, he also warrants recognition today as an underrated, forgotten pioneer in plant pathology. ***Now you know the rest of the story.***

References

Ainsworth, G. C. 1981. Introduction to the history of plant pathology. Cambridge University Press, Cambridge.

Bulloch, W. 1938. The history of bacteriology. Oxford University Press, London, 422 pp.

Estey, R. H. 1975. A note on Casimir-Joseph Davaine 1812-1882. Agricultural History 49: 549-552.

Harveson, R. M. 2015. The bacterium of many colors. APS Press, St. Paul, MN, 288 pp.

Kennedy, B. W., Widin, D. D., and Baker, I. F. 1979. Bacteria and the cause of diseases in plants: a historical perspective.

CHAPTER 48

George Clinton, Early American Plant Pathologist

Introduction

George Perkins Clinton was a highly significant figure in American phytopathology in the first two decades of the 20th century. His merit and contributions were recognized by early colleagues beyond his state service in Connecticut, but are largely forgotten today. He has been acknowledged as an authority on the cereal smuts, but his impacts reach far beyond this group of fungal diseases. His story and contributions to plant pathology are the rest of the story.

Early Life and Education

George Clinton was born in 1867 in Polo Illinois, from a homesteading, old New England family with a rich tradition for military service. Both great grandfathers served in the American Revolution. His paternal grandfather fought in the war of 1812, with another uncle participating in the Civil War.

His father was a newspaper editor so he learned early about printing, editing and writing from his experiences assisting his father. He possessed an interest in collections, which strongly correlated with his enthusiasm for botany. As a young boy, he began building a personal herbarium of both fungi and plants; a passion that continued throughout the rest of his life.

After high school, Clinton enrolled at the University of Illinois, and studied under Thomas Burrill. He received the BS in 1890 and was appointed as assistant botanist at the Illinois Agricultural Experiment Station and assistant botanist with the University of Illinois, working primarily on fungi, and earning a MS in botany in 1894. He continued with Burrill until 1900, when he enrolled at Harvard as a graduate student working with Harvard's eminent mycologists, William Farlow and Roland Thaxter. He was awarded another MS in 1901 and the Doctor of Science degree in 1902. After completing the doctorate, he became the botanist at the Connecticut Agricultural Experiment Station replacing the previously departed Thaxter.

Significant Achievements

In his role as botanist in the Experiment Station, Clinton worked substantially on pathogenic fungi with a particular interest in the rusts and smuts. However, his research was eclectic, extending into many other diseases and pathogens. Diseases of potato and tobacco (wildfire and black root rot) were targets of early investigations. He was also known for research on forestry diseases such as Chestnut blight, Dutch Elm Disease and white pine blister rust as well some of the first work with the filterable mosaic diseases (viruses) in the early 1900s.

Specific examples of his achievements include his outstanding studies with potato late blight. He was the first to identify and create oospores of the late blight pathogen (*Phytophthora infestans*) in culture utilizing oatmeal agar. This discovery was instrumental in clearing pre-existing, dubious concepts and experimentally acquitting Anton de Bary's work demonstrating that fungi could be plant pathogens.

He began virus investigations in 1903, demonstrating to growers the importance of careful sanitation in preventing the spread of tobacco mosaic. It was determined that the disease was easily transmitted mechanically by rubbing sap from infected leaves onto healthy plants. He furthermore discovered that this same virus was also the cause of the tomato mosaic disease.

Clinton, with assistant Florence McCormick, utilized a leaf assay for studying white pine blister rust. This was a method McCormick had created for artificial infections in petri dishes on detached leaves, which kept obligate pathogens alive for weeks. It was with this method that she and Clinton also discovered that the

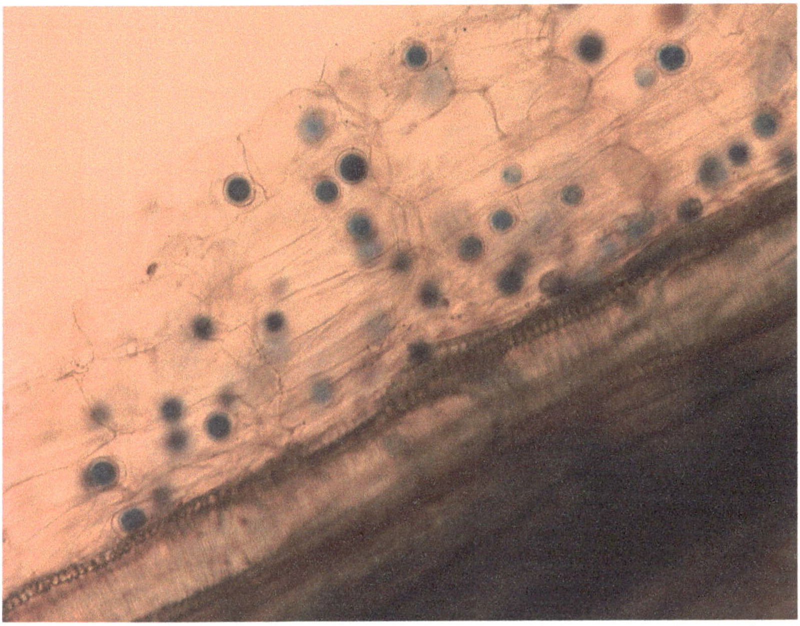

Oospores characteristic of the late blight causal agent (*Phytophthora infestans*) within plant tissues. The fungus overwinters and is infective from these structures in soil. *Photo credit: R. M. Harveson*

blister rust pathogen entered pine trees through leaf stomata.

Personality

By all accounts, Clinton was described as a modest and unassuming man that sincerely cared about helping others, expressing almost embarrassment with accepting awards. He was never too busy to address requests selflessly.

He also had a personal interest in grower problems, preferring to conduct experiments in the grower's field where the diseases had been observed, and was an early advocate of establishing working relationships with the growers.

Admiration and respect by his peers, resulted in a constant demand for his counsel. This is readily illustrated with his association with numerous scientific organizations. He was a fellow of the American Academy of Arts and Sciences, the National Academy of

Sciences, and the APS, serving as President in 1912, as well as an active member of almost a dozen other societies.

In 1908, Harvard called on him to study fungal diseases of the brown tail moth. He also lectured in forest pathology for 12 years and served as a botanical research associate at nearby Yale University from 1926-1929.

He often was called upon to travel to distant places to help solve various problems. He was sent to Puerto Rico in 1904 by the USDA to study and investigate coffee rust, and in 1909 was requested and financed by Harvard to visit Japan to collect potential parasites for controlling the gypsy moth. He traveled all over U.S., Canada, Central America and the Caribbean, as well as three trips to Europe. However, no matter where he went, he was constantly collecting both fungi and plants.

Clinton also developed and planned the library at the experiment station obtaining foreign and domestic journals and rare books. Upon retirement he bequeathed all his books, bulletins, papers, and letters to the station, including his personal herbarium species.

His only child, a son, was killed in France during the First World War, an event that plagued him the rest of his life. He never fully recovered, and his only solace and escape was to continue working exceptionally hard.

Retirement

Clinton maintained his position with the Connecticut Agricultural Experiment Station for 35 years until retiring in July 1937. However, after retirement he was still retained voluntarily as a consulting botanist on the staff of the station. Each morning he was found routinely working at his desk. He died suddenly on August 13, 1937 after over-exertion on a hot day. A few days before becoming sick, he spent an entire day in potato fields surveying for the potato late blight disease, which is thought to have hastened his death. ***Now you know the rest of the story.***

References

East, E. M. 1938. George Perkins Clinton (1867-1937). Proceedings of the American Academy of Arts and Sciences. 72: 354-358.

Maloy, O. C. 2001. George Perkins Clinton. Pages 227-228 in Encyclopedia of Plant Pathology. vol 1. O. C. Maloy and T. D. Murray, eds. John Wiley and sons, Inc. New York, 1346 pp.

McCormick, F. A. 1938. George Perkins Clinton (1867-1937). Phytopathology 28: 379-387.

Thom, C., and East, E. M. 1938. Biographical memoir of George Perkins Clinton. National Academy of Sciences 20: 181-196.